# Spatial
# Data Quality

# Spatial
# Data Quality
## From Process to Decisions

Edited by
Rodolphe Devillers
Helen Goodchild

CRC Press
Taylor & Francis Group
Boca Raton London New York

CRC Press is an imprint of the
Taylor & Francis Group, an **informa** business

CRC Press
Taylor & Francis Group
6000 Broken Sound Parkway NW, Suite 300
Boca Raton, FL 33487-2742

First issued in paperback 2017

© 2010 by Taylor and Francis Group, LLC
CRC Press is an imprint of Taylor & Francis Group, an Informa business

No claim to original U.S. Government works

ISBN-13: 978-1-4398-1012-5 (hbk)
ISBN-13: 978-1-138-11782-2 (pbk)

**Visit the Taylor & Francis Web site at**
**http://www.taylorandfrancis.com**

**and the CRC Press Web site at**
**http://www.crcpress.com**

# Contents

## SECTION III  IMAGERY

## SECTION IV  LEGAL ASPECTS

# Contributing Authors

**Ali A. Alesheikh**
Faculty of Geomatics Engineering
K.N. Toosi University of Technology
Tehran, Iran
alesheikh@kntu.ac.ir

**Abbas Alimohammadi**
Faculty of Geomatics Engineering
K.N. Toosi University of Technology
Tehran, Iran
abb_alimoh@kntu.ac.ir

**Yvan Bédard**
Department of Geomatic Sciences and
   Centre for Research in Geomatics
Université Laval
Québec, QC, Canada
yvan.bedard@ulaval.ca

**Lotfi Bejaoui**
Department of Geomatic Sciences and
   Centre for Research in Geomatics
Université Laval
Québec, QC, Canada
lotfi.bejaoui.1@ulaval.ca

**Rick L. Bello**
Department of Geography
York University
Toronto, ON, Canada
bello@yorku.ca

**Debajyoti Bhowmick**
Department of Earth Observation Science
International Institute for Geo-Information
   and Earth Observation
Enschede, The Netherlands
bhowmick@itc.nl

**Jean Brodeur**
Center for Topographic Information
Natural Resources Canada
Sherbrooke, QC, Canada
brodeur@nrcan.gc.ca

**Petra Budde**
Department of Earth Observation Science
International Institute for Geo-Information
   Science and Earth Observation
Enschede, The Netherlands
budde@itc.nl

**Mario Caetano**
Remote Sensing Unit
Portuguese Geographic Institute
Lisboa, Portugal
mario.caetano@igeo.pt

**Lisa M. Campbell**
Office of the Privacy Commissioner of
   Canada
Ottawa, ON, Canada
lcampbell@privcom.gc.ca

**Cláudio Carneiro**
Geographical Information Systems Laboratory
Ecole Polytechnique Fédérale de Lausanne
Lausanne, Switzerland
claudio.carneiro@epfl.ch

**Jennifer A. Chandler**
Common Law Section, Faculty of Law
University of Ottawa
Ottawa, ON, Canada
chandler@uottawa.ca

**Alexis Comber**
Department of Geography
University of Leicester
Leicester, United Kingdom
ajc36@le.ac.uk

**Mamoud Reza Delavar**
GIS Division, Department of Surveying
   and Geomatics Engineering
College of Engineering, University of
   Tehran
Tehran, Iran
mdelavar@ut.ac.ir

**Carolyn H. Eyles**
School of Geography and Earth Sciences
McMaster University
Hamilton, ON, Canada
eylesc@mcmaster.ca

**Rainer Feucht**
Vermessung Dipl. Ing. Erich Brezovsky
Gänserndorf, Austria
rainer.feucht@gmx.at

**Yevgeniya Filippovska**
Institute for Photogrammetry
Universität Stuttgart
Stuttgart, Germany
yevgeniya.filippovska@ifp.uni-stuttgart.de

**Peter Fisher**
Department of Geography
University of Leicester
Leicester, United Kingdom
pff1@le.ac.uk

**Cidália C. Fonte**
Department of Mathematics
University of Coimbra
Coimbra, Portugal
cfonte@mat.uc.pt

**Andrew U. Frank**
Institute for Geoinformation and
  Cartography
Vienna University of Technology
Vienna, Austria
frank@geoinfo.tuwien.ac.at

**François Golay**
Geographical Information Systems
  Laboratory
Ecole Polytechnique Fédérale de Lausanne
Lausanne, Switzerland
francois.golay@epfl.ch

**Luísa M.S. Gonçalves**
Civil Engineering Department
Polytechnic Institute of Leiria
Leiria, Portugal
luisag@estg.ipleiria.pt

**Nicholas A.S. Hamm**
Department of Earth Observation Science
International Institute for Geo-Information
  and Earth Observation
Enschede, The Netherlands
hamm@itc.nl

**Tom Hoogland**
Soil Science Centre
Alterra, Wageningen University and
  Research Centre
Wageningen, The Netherlands
tom.hoogland@wur.nl

**Farhad Hosseinali**
Faculty of Geomatics Engineering
K.N. Toosi University of Technology
Tehran, Iran
frdhal@gmail.com

**Eduardo N.B.S. Júlio**
Department of Civil Engineering
University of Coimbra
Coimbra, Portugal
ejulio@dec.uc.pt

**Martin Kada**
Institute for Photogrammetry
Universität Stuttgart
Stuttgart, Germany
martin.kada@ifp.uni-stuttgart.de

**Yogesh Kant**
Indian Institute of Remote Sensing
Government of India
Dehradun, Uttarakhand, India
yogesh@iirs.gov.in

**Mina Karzand**
Audiovisual Communications Laboratory
Ecole Polytechnique Fédérale de Lausanne
Lausanne, Switzerland
mina.karzand@epfl.ch

**Arie van Kekem**
Soil Science Centre
Alterra, Wageningen University and
  Research Centre
Wageningen, The Netherlands
arie.vankekem@wur.nl

**Martin Knotters**
Soil Science Centre
Alterra, Wageningen University and
    Research Centre
Wageningen, The Netherlands
martin.knotters@wur.nl

**Connie Ko**
Department of Geography
York University
Toronto, ON, Canada
cko@yorku.ca

**Yue M. Lu**
Audiovisual Communications Laboratory
Ecole Polytechnique Fédérale de Lausanne
Lausanne, Switzerland
yue.lu@epfl.ch

**Kelsey MacCormack**
School of Geography and Earth Sciences
McMaster University
Hamilton, ON, Canada
maccorke@mcmaster.ca

**Haixia Mao**
Department of Land Surveying and Geo-
    informatics
The Hong Kong Polytechnic University
Hung Hom, Kowloon, Hong Kong
jessie.mao@polyu.edu.hk

**Edward J. Milton**
School of Geography
University of Southampton
Southampton, United Kingdom
ejm@soton.ac.uk

**Hamid Kiavarz Moghaddam**
Department of Geomatics Engineering
Tehran University
Tehran, Iran
hkiavarz@gmail.com

**Mir Abolfazl Mostafavi**
Department of Geomatics Sciences and
    Center for Research in Geomatics
Laval University
Québec, QC, Canada
mir-abolfazl.mostafavi@scg.ulaval.ca

**Gerhard Navratil**
Institute for Geoinformation and
    Cartography
Vienna University of Technology
Vienna, Austria
navratil@geoinfo.tuwien.ac.at

**Harlan J. Onsrud**
Department of Spatial Information Science
    and Engineering
University of Maine
Orono, ME, U.S.A.
onsrud@spatial.maine.edu

**Michael Peter**
Institute for Photogrammetry
Universität Stuttgart, Germany
michael.peter@ifp.uni-stuttgart.de

**Bryan C. Pijanowski**
Department of Forestry and Natural
    Resources
Purdue University
West Lafayette, IN, USA
bpijanow@purdue.edu

**François Pinet**
Cemagref
Clermont-Ferrand, France
francois.pinet@cemagref.fr

**Rahul Raj**
Indian Institute of Remote Sensing
Government of India
Dehradun, Uttarakhand, India
rahulosho@gmail.com

**Tarmo K. Remmel**
Department of Geography
York University
Toronto, ON, Canada
remmelt@yorku.ca

**Mehrdad Salehi**
Department of Geomatic Sciences and
    Centre for Research in Geomatics
Université Laval
Québec, QC, Canada
merhdad.salehi.1@ulaval.ca

**Hadis Samadi Alinia**
GIS Division, Department of Surveying
  and Geomatics Engineering
College of Engineering, University of
  Tehran
Tehran, Iran
alinia@ut.ac.ir

**Farhad Samadzadegan**
Department of Geomatics Engineering
Tehran University
Tehran, Iran
samadz@ut.ac.ir

**Tarek Sboui**
Department of Geomatic Sciences and
  Centre for Research in Geomatics
Université Laval
Québec, QC, Canada
tarek.sboui.1@ulaval.ca

**Teresa Scassa**
Common Law Section, Faculty of Law
University of Ottawa
Ottawa, ON, Canada
teresa.scassa@uottawa.ca

**Michel Schneider**
LIMOS
Université Blaise Pascal
Aubière, France
schneidr@isima.fr

**Wenzhong Shi**
Department of Land Surveying and Geo-
  informatics
The Hong Kong Polytechnic University
Hung Hom, Kowloon, Hong Kong
lswzshi@polyu.edu.hk

**Alfred Stein**
Department of Earth Observation Science
International Institute for Geo-Information
  Science and Earth Observation
Enschede, The Netherlands
stein@itc.nl

**Amin Tayyebi**
GIS Division, Department of Surveying
  and Geomatics Engineering
College of Engineering, University of Tehran
Tehran, Iran
amin.tayyebi@gmail.com

**Yan Tian**
Department of Electronic and Information
  Engineering
Huazhong University of Science and
  Technology
Wuhan, China
tianyan2000@126.com

**Valentyn Tolpekin**
Department of Earth Observation Science
International Institute for Geo-Information
  Science and Earth Observation
Enschede, The Netherlands
tolpekin@itc.nl

**Martin Vetterli**
Audiovisual Communications Laboratory
Ecole Polytechnique Fédérale de Lausanne
Lausanne, Switzerland
martin.vetterli@epfl.ch

**Henk Vroon**
Soil Science Centre
Alterra, Wageningen University and
  Research Centre
Wageningen, The Netherlands
henk.vroon@wur.nl

**Richard Wadsworth**
Centre for Ecology and Hydrology -
  Monks Wood
Huntingdon, United Kingdom
rawad@ceh.ac.uk

**Gwen Wilke**
Institute for Geoinformation and Cartography
Vienna University of Technology
Vienna, Austria
wilke@geoinfo.tuwien.ac.at

**Mohammad Javad Yazdanpanah**
School of Electrical and Computer
  Engineering
College of Engineering, University of
  Tehran
Tehran, Iran
yazdan@ut.ac.ir

**Mamushet Zewuge Yifru**
Department of Earth Observation Science
International Institute for Geo-Information
  Science and Earth Observation
Enschede, The Netherlands
yifru08938@alumni.itc.nl

# Introduction

Producing maps that depict the real world accurately has been a concern of cartographers for centuries. This concern still exists in 'modern' cartography, and the field of Spatial Data Quality (SDQ) has been, since the early years of GIS, one of the core sub-disciplines of the Geographic Information Sciences. The increasing awareness by the scientific community of spatial data quality issues led, in 1999, to the organisation of the first International Symposium on Spatial Data Quality (ISSDQ) by Dr. John Shi in Hong Kong. Ten years later, this book presents the contributions from the 6th ISSDQ conference, held in St. John's, Newfoundland and Labrador, Canada, from July 5th to 8th 2009, and reflects changes in practice in response to the rapid technological developments of the previous decade.

The theme selected for the conference and used for the title of this book is *Spatial Data Quality: From Process to Decisions*. It attempts to draw the attention of the SDQ community and, more generally, of the GISciences/Geomatics community, to a long-term trend in the SDQ research agenda – that spatial data quality goes beyond solely data accuracy concerns. In today's world, geospatial data are used widely to support decisions of all types. An increasing importance is given to the relationships that exist between the quality of spatial data used and the quality of the decisions being made using these data. This includes, for instance, semantic/ontological aspects (see Chapters 1 to 3 of this book), and raising the awareness of end-users about data quality and uncertainty. While some of these issues have a long history of scholarship, the increasing importance of the Web, particularly in widening the distribution of existing geospatial datasets, has meant that the likelihood of people using datasets of inappropriate quality for a given task is greatly increased. The burgeoning access to geospatial data by the general public (e.g., Google Earth, Google Maps), and the subsequent increase in user-generated content provided by the 'Web 2.0 generation', have significantly altered the typical process that was previously used to produce, distribute and use geospatial data. This book features a number of fundamental and applied contributions related to SDQ with, among other aspects, papers discussing legal issues related to spatial data quality that have arisen due to this widespread availability of spatial data (see Section IV).

Some of the main challenges that the SDQ community faces are related to the fact that each paper or digital map is only one of numerous possible representations of the real world. It is accepted that maps are merely a model created in a given context for a specific purpose. Maps look at a subset of the world, at a certain scale, with certain rules, using a certain language for the representation and descriptions of objects. Hence, there cannot be such a thing as a map that would perfectly represent the world as, by their nature, maps deliberately and necessarily omit or simplify phenomena from the real world in order to improve the communication process and fit the constraints imposed by computer and GIS software. If maps recorded the world in its overall complexity, the task of the SDQ community would be much easier, as it could focus on ways to quantify or qualify the differences between map features and the world's phenomena (although several aspects of this task are also problematic). Maps are models, however, and as such we need to understand how these models simplify the reality, how scale plays a role in the abstraction (e.g., Chapter 2), how people decide to name and classify real-world phenomena (e.g., Chapter 1) and, finally, how these maps fit the needs of users that may use them to make various types of decisions - the concept of *fitness for use* - (e.g., Chapter 11).

As you might have noticed while opening this book, the map published on its cover includes a number of (deliberate) mistakes. These mistakes illustrate various types of errors that can be found on maps, the description of those errors being one of the key components of the field of spatial data quality. Standard methods of documenting such errors led to the publication of a number of international standards such as FGDC and ISO 19113/19115. The map includes objects poorly located ('positional accuracy' issues, e.g., Sri Lanka, Madagascar and a 'slight' enlargement of Newfoundland, where the conference was held), objects missing ('completeness/omission', e.g., New Zealand) or objects that do not exist in reality ('completeness/commission', e.g., the large island located in the Indian Ocean), problems with the names associated to objects ('attribute accuracy', e.g., Europe, South America), etc. If you wish to find all of the different problems, there are over 30 of them, and potentially more if you classify some errors in different quality elements.

The book is structured in four sections: the first section discusses conceptual approaches to spatial data quality; the second presents a number of applications of spatial data quality methods; the third looks at spatial data quality issues in the realm of remote sensing; and the final section presents papers that consider the interface between the law and spatial data quality. In addition to the main chapters presented in each section, a number of shorter notes present on-going or recent research projects investigating various aspects of spatial data quality.

We hope that reading this book will give you some sense of the latest research developments made in spatial data quality for the benefit of all users of geospatial data.

**Rodolphe Devillers and Helen Goodchild**

# Acknowledgements

A number of people helped with the organisation of the symposium that led to this book. We would like to thank them warmly.

On the scientific side, all the chapters published in this book were carefully reviewed by a program committee, which includes Francisco Javier Ariza López (Spain), Yvan Bédard (Canada), Nicholas Chrisman (Canada), David Coleman (Canada), Alexis Comber (UK), Sytze de Bruin (The Netherlands), Mahmoud Reza Delavar (Iran), Hande Demirel (Turkey), Rodolphe Devillers (Canada), Matt Duckham (Australia), Peter Fisher (UK), Andrew Frank (Austria), Marc Gervais (Canada), Michael Goodchild (USA), Gerard Heuvelink (The Netherlands), Robert Jeansoulin (France), Werner Kuhn (Germany), Tarmo Remmel (Canada), Sylvie Servigne (France), John Shi (Hong Kong), Alvin Simms (Canada), Alfred Stein (The Netherlands), and Kirsi Virrantaus (Finland). Reviewing papers is time consuming, not always recognised as it should, but critically important for the quality of the research done in our scientific community. Thank you for taking the time to provide constructive comments on each contribution.

Locally, a number of people contributed to organizing the conference, including the members of the local organizing committee: Randy Gillespie (Marine Institute), Alvin Simms, Helen Goodchild, Randal Greene and Krista Jones (MUN Geography). Harriett Taylor, Charles Mather and Charlie Conway (MUN Geography) and Sherry Power (Marine Institute) also helped with different parts of the organisation. And many thanks go to the students that volunteered to help during the event!

Thank you also to our sponsors that helped support part of the financial cost related to the conference, including the Canadian GEOIDE Network, GeoConnections, Memorial University of Newfoundland, Compusult, the Newfoundland Branch of the Canadian Institute of Geomatics, and the Atlantic Canada Opportunities Agency (ACOA). The International Society of Photogrammetry and Remote Sensing (ISPRS) are also thanks for having agreed to jointly organize this event.

Finally, our thanks go to the editorial team of CRC Press/Taylor & Francis that worked under pressure to deliver this book within a short timeframe.

# SECTION I – CONCEPTUAL APPROACHES TO SPATIAL DATA QUALITY

# 1

# What's in a Name? Semantics, Standards and Data Quality

*Peter Fisher, Alexis Comber & Richard Wadsworth*

> "I don't know what you mean by 'glory,'" Alice said.
> Humpty Dumpty smiled contemptuously. "Of course you don't – till I tell you. I meant 'there's a nice knock-down argument for you!'"
> "But 'glory' doesn't mean 'a nice knock-down argument,'" Alice objected.
> "When I use a word," Humpty Dumpty said, in a rather scornful tone, "it means just what I choose it to mean – neither more nor less."
> "The question is," said Alice, "whether you can make words mean so many different things."
>
> Lewis Carroll, 1871
> *Through the Looking-Glass, and what Alice found there*

## Abstract

*Most formal standards (ISO, FGDC, ANZLIC, Dublin Core, etc.) are concerned (in practice) with the discovery and acquisition of digital resources (their location, authorship, access, etc.). When applied to descriptions of geographical data and information they usually ignore the semantic description of the content of the data. Description is usually restricted to the format, geometry and information production, rather than the meaning. In many senses, however, the semantics embedded in the information by human interpretation of a phenomenon are crucial to the users of the information and deserve to be treated more comprehensively. This paper reviews the importance of semantics and its treatment within current metadata and data quality standards. Thorough consideration of semantics (as part of standards or not) would provide a background to help users understand the information and its relevance to their application. The potential for a lack of user understanding has emerged through the increasingly wide access to and use of spatial information in GRID computing and spatial data infrastructures by users and communities without any prior knowledge of the information. Potential and current users may lack any background in the domain which the information is describing, and have a total separation from the producers of the information. Clearly the best solution to the problem identified would be a complete disciplinary understanding of all information domains a user might be employing, as this is practically impossible we suggest new types of metadata and some novel tools to assist users in assessing the spatial information relative to their intended application.*

## 1.1  Introduction

In this paper we document the treatment of data semantics in different data and metadata standards. At its simplest, semantics is the study of meanings, and we wish to restrict it to meaning in language as opposed to meaning in any other semiotic devices (Merriam-Webster, 2009). We do not specifically address data quality, believing that semantics are so central to the understanding and use of information that they have a central role in data quality. Our primary concern is to consider the problem of divergent semantics (data-to-data or application-to-data) from the perspective of the data user. The arguments we present are a confluence of ideas and arguments that have grown out of a number of activities (Comber *et al.*, 2005a; 2006; 2008, among others). Not all of the ideas presented are novel but we believe that bringing them together within a critique of current metadata and data standards specifications offers insight into how semantics are currently addressed and how they might be better accommodated in the future.

Current metadata paradigms do not appear effective in replacing the old-fashioned and sometimes informal interaction between the provider and user of data. Instead they reflect the position where it is assumed that computers "replace the extended and often confused process by which we learn the meanings of terms and languages with precise, instantaneous translators" (Goodchild, 2006: 690). Comber *et al* (2005a) examined the issues relating to alternative conceptions of spatial information using the example of land cover. The lack of meta-information about the meaning and semantics embedded in a dataset and its classification is problematic for users unfamiliar with the data.

Historically, if users wished to obtain data they negotiated with the producers: producers were concerned about inappropriate use of "their" data and acted as gatekeepers. Now users are able to download data from the internet and facilitated by Spatial Data Infrastructures (SDIs) such as INSPIRE without going through any "gatekeeper", increasing the separation of data providers and users. With little meta-information about embedded data concepts and the assumptions that underpin them, users are encouraged to assume that it is appropriate for their analyses, despite the potential for huge mis-matches between what has been called 'User-Producer' and 'Product-Problem' ontologies (Van de Vlag *et al.*, 2005; Shortridge *et al.*, 2006).

The objective of this paper is to provide an overview of the way that current metadata and standards for Geographical Information (GI) deal with data semantics. The apparent assumption has been that any users of the information will either be familiar with the nature of the information or be prepared to invest time and effort in acquiring familiarity, or, more likely, that the knowledge will be "self-evident". The usefulness of standards will not be realized unless the legitimate differences in meaning and methods between disciplines are acknowledged.

## 1.2  Data Semantics

Understanding the embedded and explicit semantics of a dataset is critical for user assessments of fitness for use (its quality). Within GI, descriptions of objects and object classes may be composed of a single word (Building, Tree), a short phrase (coniferous forest, upland moors, etc.) or longer textual descriptions. Data semantics also include the general description of a dataset and its characteristics and limitations, including, in a few cases, what is explicitly omitted (e.g. Ordnance Survey, 2009). Semantics are concerned with the meaning in language of those descriptions (Calvani, 2004). They are critical to understanding what a description means, to filling out the description of classes and for providing justifications for measurements.

Many valuable aspects of GI are embedded in the semantic descriptions and there is the potential for confusion in a world of heterogeneous semantics where many words are used for the same thing and the same thing is described in many ways. GI is created by human interpretation of observations or measurements (Comber *et al.*, 2005b). The semantic information therefore embeds much knowledge of how the interpretation is made and the criteria that go into that interpretation. Often this knowledge is specific to the subject domain of the information. Some of this knowledge relates to class allocation, conversion and inclusion and to boundary conditions that may be applied within a particular discipline. The project brief and socio-political context have considerable impact on the mapped characteristics of a database, and on the basis for naming classes and what those class names imply (Martin, 2000; Comber *et al.*, 2003). Together, this is the semantic/socio-scientific background to the mapped information, and understanding of these aspects of disciplinary knowledge is essential to the effective and efficient use of the spatial information. A typical result of ignoring this sort of knowledge is rediscovery of relationships which are actually created in the mapping, such as perhaps geological boundaries coinciding with soil boundaries (because mapping of one is used to derive the map of the other).

The real world is infinitely complex and all representations (such as maps) involve intellectual processes such as abstraction, classification, aggregation, and simplification. Natural resource inventories of the same phenomenon at different times or by different people may vary for a number of reasons that have nothing to do with changes or differences in the feature being mapped; but reflect advances in data collection technology or background science. Similarly collectors of social information and demographic statistics make decisions about the information to collect, and different collection agencies (national census offices, opinion pollsters or researchers) may collect different socio-economic variables to measure the same property of their target population. Often these differences in representation are known amongst specialists but are not always explicitly acknowledged. Thus, in the creation of any GI there are a series of choices about what to map and how to map it which, at least, depends on:

- The commissioning context, specifically legislation and policy;
- The method of information measurement;
- Observer variation;
- Institutional variation in classes, variables and definitions;
- The varied training of all the different scientific staff from supervisor to surveyor; and
- The intellectual process of transforming the observations into a "map".

These are in addition to issues relating to the survey process, such as scale and the existence of base mapping, and to improvements in technology where revised procedures can be perceived to offer improved insight about the phenomenon. The point is that if two observers working for the same organization are sent to collect information for some purpose over a single area, they may measure different properties and may come up with different mappings (Cherrill and McLean, 1999). With willingness and communication they will negotiate those differences to come up with a consistent mapping. If they work for different organizations the differences may be more profound and there may be no opportunity to resolve them. The net result can have a profound effect on the final spatial data product and the meaning of the information in its widest sense: different choices result in different representations and the variability between different, but equally valid, mappings of the same real world objects (Harvey and Chrisman, 1998).

In summary, over time scientific knowledge advances, policy objectives change and technology develops. These contribute to a common problem in the environmental and social sciences where change may be apparent even when phenomena are not expected to change (and did not). Most surveys therefore establish a new "base line" even if they are part of a sequence of surveys, repeatedly describing the same features of interest. Whether the mapping is of solid geology which may not be expected to change, or land cover which is expected to show some change, it is not always possible to tell whether differences between "maps" represent changes in understanding, technology, objectives or in the phenomenon forming the subject of the map. The changes observed may be significant and interesting, or they may be the product of the process of creating the information. For users unfamiliar with the information it may be hard or impossible to understand the differences in meaning.

## 1.3 Semantics in Metadata

### 1.3.1 Metadata and Standards

Put simply, metadata is

"Data about data or a service. Metadata is the documentation of data. In human-readable form, it has primarily been used as information to enable the manager or user to understand, compare and interchange the content of the described data set" (ISO, 2003a).

It is therefore clear that semantics of a dataset are a legitimate area which might be considered by metadata.

Arising from the desire to transfer GI between users and the wish to find out the existence of data conveniently, the GI community developed metadata standards. Initially these were included in the FGDC Spatial Data Transfer Standards (DCDSTF, 1988), but as the number of possible metadata fields grew, separate standards arose. The aims were:

- To allow individuals within an organization to discover the existence of any dataset of interest and how to acquire it;
- To help organize and maintain an organization's internal investment in spatial data;
- To provide information about an organization's data holdings to data catalogues, clearinghouses, and brokerages; and

- To provide information to process and interpret information received from an external source.

However, the only semantic information referred to in these standards and many others that have followed (such as FGDC, 1998; ANZLIC, 2001; ISO, 2003a, 2003b) is the list of data categories, and these may be considered as "optimal" rather than "required".

The inclusion of data for ready access over massively networked computer resources (SDIs and the GRID) has led to further concern with metadata. For example, the NERC DataGrid in the UK has specified five types of metadata or 'elements' (Lawrence et al., 2003):

A.  Usage metadata to provide information about the data needed by the processing and visualisation services;
B.  Generic complete metadata to provide high-level statements of entities and the relations between them and to summarise the data model in a user friendly form;
C.  Metadata generated to describe documentation and annotation;
D.  Discovery metadata to be created by the data producer, conforming to ISO19115 (ISO 2003a);
E.  Extra discipline specific information.

In this framework, communities of users identify portions of the metadata in which they have personal expertise, populating slots or elements where necessary under element E. This view of data and user communities is similar to that put forward by the Open Geospatial Consortium (OGC) (OGC refers to them as 'information communities'; www.opengeospatial.com). However, if one accepts that it is possible to identify various user communities, metadata for a single data set may have to be populated $N$ times for the $N$ different communities of data users. Quite apart from the effort involved in creating this many different versions of the metadata, there is the problem of personal identification with any given community: 'to which communities do I think I belong?', 'will community $X$ take me seriously?'. Furthermore, use-communities can be diverse and are difficult to define and identify, and Visser et al. (2002) note that communication between different information communities can be very difficult because they do not agree on common conceptualizations.

Some user-communities are well defined and well established. They develop their own standards and discuss a range of advanced concerns. For example the Climate and Forecast Metadata Convention community (Eason et al., 2009) lists many discussion threads, some of which are trying to get technical agreement on a physical measurement, whilst others are concerned with what to include or exclude (e.g. compare the 2006 threads for "omega" and the "shape of the earth").

There is no recognition within current metadata that the specification of the conceptual model may vary and needs to be shared. Implicit in the received view in metadata standards is the idea that identifying the 'universe of discourse' within communities is unproblematic. The assumption seems to be that from an agreed "universe of discourse" encoding protocols, application schemas, conceptual modelling and feature catalogues – each with their own ISO standard – can be developed as a process for within community governance of semantics. However, defining a universe consistently and with consensus is problematic for many within the GI and related communities. There is very little agreement on the extent of the universe of discourse. Furthermore, the metadata standard for one community may be meaningless to another community although the information described may be very suitable or even important for applications of both; the communities may be separated by their semantic heterogeneity.

### 1.3.2  Metadata and fitness for use

Standards for metadata are useful because in theory they provide a common language, enabling parties to exchange data without misunderstandings. However their specification (content) is always a compromise and necessarily lags behind research activity and sometimes industrial practice. For example, a recent book on spatial data standards took ten years from inception to being published (Moellering, 2005). Current metadata standards are characterized as being static and grounded in the production of spatial data and in the idea of objective observations of spatial information. This is a poor model for how most spatial information is observed, and as a consequence metadata standards do not represent the depth of knowledge held within a scientific community.

More pertinently metadata standards in general, and data quality as a portion of those standards, have a number of characteristics. First, their cartographic legacy reflects data production and the stages involved in data creation. Methods and data sources define *lineage* and assessment of results defines *accuracy, consistency* and *completeness* (Fisher, 1998). These historical and cartographic legacies are seen in the way that metadata reports the degree to which database objects generalize complex information for the purposes of display (Fisher, 1998). Second, they record passive attributes, characterized as the easily measurable aspects of data (Comber *et al.*, 2005b), which cannot easily be interpreted or activated within potential applications (Hunter, 2001). Third, as currently implemented they generally provide overall or global measures of data quality, rather than ones that relate to individual map objects, even though those are enabled in the standards. Fourth, the measures are a signal of an accepted paradigm that the producer has put in the required effort. The metadata measures can also be seen in the context of "economic game theory" as the user has imperfect knowledge of how much effort the producer puts into data quality reporting. The net impact is that current metadata reporting for spatial data does not give any information to the user about how to best exploit the data. In order to assess data quality (a relative concept), users need to relate their intended use of the data with the semantic and ontological concepts of the data. Salgé (1995) introduced this concept under the term "semantic accuracy", but this concept has had little impact upon either the standards process or metadata specifications (Fisher, 2003).

In spite of the declared intention that metadata assist users in defining the fitness of a dataset for their application, the standards in general and semantic descriptions in particular, are not easy to relate to intended use. Rather, the descriptions reflect data production interests showing that the data producer can follow a recipe to within particular tolerances (Comber *et al.*, 2005b). They do not attempt to communicate the producer's full knowledge of the data nor to facilitate the role of the user. As an example, the recent INSPIRE draft rules for metadata (INSPIRE, 2007: 17) explicitly state:

> "Attempts to objectively rate (and publish in metadata) the "usefulness" of a service, such as that it produces correct responses or behaviours, will almost certainly create problems among service vendors, and would likely do more harm than good to consumers. Most other markets rely on informal user feedback as the ultimate test as to whether or not a product or service is useful, a good value, etc. This feedback appears spontaneously in news and mail forums, in the popular press, and by word-of-mouth"

The implication is that it is not the role of metadata to help users make fitness assessments despite that as originally conceived and commonly understood, metadata should provide something other than discovery information; when those involved in metadata specifications reveal their hand, only lip-service is paid to the user.

### 1.3.3 Standards and semantics

Within the standards for spatial information, data semantics are supported in many different ways. In some instances semantics are embedded in the metadata standards calling for a short phrase for each mapping unit; ISO 19115, for example, calls for a Feature Type list and spatial attribute supplement information (both of which are mandatory; ISO, 2003a). In other instances, there are attempts to specify descriptions of mapping units such as ISO 19110 (ISO, 2001) which requires the full feature catalogue. Kresse and Fadaie (2004: 38) point out that "ISO does not standardise feature catalogues and data models". The OGC Observations and Measurements specification (OGC, 2007) explicitly states that it is not concerned with semantic definition of phenomena, and that a schema for semantic definitions of property-types is beyond its scope. OGC notes that such a schema depends on shared concepts described in natural language and specification is concerned with observed properties of feature types, although the observation-model "takes a user-centric viewpoint, emphasizing the semantics of the feature-of-interest and its properties" (OGC, 2007: 10). They also take note of the encoding and call-out problems caused by varying semantics and multiple representations for both discovery and data fusion. In part the general avoidance of semantics in GI standards seems to be because the GI community is looking to the Ontology research community to provide standards for data catalogues, for example, through the Web Ontology Language OWL. However a true ontology is different from semantics, including as it does not only the feature classes but also their relationships (nouns and verbs). The traditional use of language to describe landscape and environmental features is rather different, including more expressive use of qualitative terms, quantitative thresholds and adjectives as well as nouns and verbs. Unfortunately, typical ontologies are as terse in their descriptions of classes as many current GI feature lists.

There is considerable development of standards for semantic descriptions in user communities. Metadata for Biological Records are developed under the general banner of Biodiversity Information Standards by the Taxonomic Databases Working Group. As part of this endeavour seventeen separate standards are under development and in use (http://www.tdwg.org/standards/). Many of these are either explicitly associated with spatial information, or relate to descriptions of phenomena which might be the subject of mapping. Some examples of standards from this group include:

- The Darwin Core (DwC) (DarwinCore, 2008)
- The "Structure of Descriptive Data" (SDD); and
- The "World Geographical Scheme for Recording Plant Distributions" (Brummitt, 2001).

Within the FGDC community there are also a number of identifiable user communities producing metadata standards for attribute information, at least one of which is more than 25 years old, including:

- Standards for biological metadata (FGDC, 1999), and the vegetation classification standards (FGDC, 1997a);
- Soil information standard (FGDC, 1997b);
- The metadata standard for shoreline information (FGDC, 2001);
- A standard for cadastral mapping (FGDC, 2003); and
- Standards for wetland classification (Cowardin *et al.*, 1979).

These standards present a mixture of approaches. At one extreme is that which specifies the units which will be recognised (the vegetation classification, FGDC, 1997a; the wetland standard, Cowardin *et al.*, 1979). On the other hand, the soil standard (FGDC, 1997b) presents a more generic statement of how soil information should be reported, rather than how it should be classified and surveyed, which is amply covered in other publications (Soil Survey Staff, 1999).

Within the ISO community one movement for semantic standards has come from the land cover mapping community where the UN Land Cover Classification System (LCCS; Di Gregorio and Jansen, 2000) has been submitted as a possible standard of the ISO 19000 series (Herold *et al.*, 2006). The LCCS presents a standard for land cover legends which includes literally thousands of classes, with multiple criteria diagnostic of classes and associated information. Herold *et al.* (2006) state that land cover mapping must move towards compliance with standards so as to achieve efficiency and effectiveness in international and interregional mapping.

The review of the semantic standards shows that in spatial data standards semantics are acknowledged but barely treated. Where they exist they are domain specific with little commonality between domains: they all specify different things. Soil standards are based on the history of soil mapping and identify what should be measured on soils and how that should be included in the digital record. They are independent of the Soil Taxonomy which describes how the soils should be classified for mapping. Whilst in parallel to the soil standards the vegetation mapping standards produced by the Biological Survey of the USGS (FGDC, 1997a) specify the classes that each quadrat may be called; it seems that there is no commonality between approaches in different domains.

## 1.4   Recommendations

A more user-focused definition of metadata might be:

*Information that helps users assess the usefulness of a dataset relative to their problem.*

In this definition metadata is not static information relating to the "usability" of the data, but is concerned with whether the data is "useful" for the task in hand. In our view, metadata should include information that helps users to:

- Assess the suitability of data for their use;
- Integrate the data (spatially and thematically); and
- Understand the limitations and uncertainties of any integrating activities.

### 1.4.1   Socio-political context of data creation: actors and their influence

This documents the legitimising activity, identifies the major actors in commissioning the data including data producers and users, agencies and NGOs, and their policy requirements, and the influence they exert on the process of product specification. Comber *et al.* (2003) applied this approach, to clarify the reasons behind the discordant mappings of land cover in the UK in terms of the different socio-political contexts and their influence on data conceptualisations.

### 1.4.2   Critiques of the data: academic and industrial papers

Academic or practical journal papers, magazine articles and technical reports which describe or critique uses of a dataset in particular contexts are a form of metadata. Users who wish to identify the suitability of any particular dataset for their problem would find it helpful to be directed to these papers as they provide an independent opinion of the data quality and fitness.

Metadata reporting of this type was, indeed, foreseen at the committee stage of the Proposed Standard for Digital Cartographic Data (DCDSTF, 1988). Chrisman reports that he advised on the need to include within the specification of data quality the dynamic reporting of users' experience with data in applications, but this recommendation was omitted from the final Spatial Data Transfer Standards specification (FGDC, 1998); an omission which has been propagated through most standards specifications since (Nicholas Chrisman, *pers comm.*).

### 1.4.3   Data producers' opinions: class separability

The opinions of the data producers on the separability of classes can allow informed assessments of usefulness of data to be made. Comber *et al.* (2004) and others have applied such descriptions of internal class separability as weights for assessing internal data inconsistency.

### 1.4.4   Expert opinions: relations to other datasets

Experts familiar with the data can provide measures of how well the concepts or classes in one dataset relate to those of another. This generates measures of (external) data inconsistency which can be used as weights for applications. Comber *et al.* (2004) applied this approach to determine whether variations between different datasets were due to data inconsistencies (i.e. alternative specifications) or due to actual changes in the features being recorded.

### 1.4.5  Experiential metadata

Users could provide feedback about their experiences of using the data. This could be from an application or disciplinary perspective in order to describe positive and negative experiences in using the data, and could include quantitative evaluation of semantic aspects of data. Slots (elements) in metadata taxonomies exist to hold such information, but the mechanism to allow post-production reporting of metadata does not. One possible web-based solution is to host a metadata wiki for a particular group of products.

### 1.4.6  Free text descriptions from producers

The existing and emerging metadata standards include elements for free text slots, including "Descriptions" in the Dublin Core, and "Generic" and "Extra" in the NERC DataGrid specifications. These are to be populated with either producer or user community perspectives but are currently not used extensively. As noted above, information describing the full complexity of a dataset used to be published in a survey memoir. In respect to the dataset, such information is metadata and is supported by further levels of metadata in survey manuals and disciplinary textbooks. Wadsworth *et al.* (2006) have concluded that free-text *descriptions* of classes, which are longer than about 100 words, can be mined for information which is helpful to users who are unfamiliar with the epistemology, ontology and semantics of the data-type or classes. Currently this information is unavailable or ignored in both standards and implementations of metadata, but could be included as generic metadata (FGDC, 1999; ISO, 2003a), as domain specific metadata and data specification (FGDC, 1997a; 1997b), or as feature catalogues (ISO, 2001). To exploit the information embedded in this text, novel approaches to metadata mining and analysis are needed, so that users unfamiliar with such text can explore the descriptions and relationships between classes (Wadsworth *et al.*, 2006).

## 1.5  Conclusions

The analysis of spatial metadata specification in standards suggests that for GI in general the treatment of semantic information has been at best partial. Where they exist, specifications are domain specific with little commonality between domains. The apparent assumption has been that users of the information will either be familiar with the nature of the information or be prepared to invest time and effort in acquiring familiarity, or that the knowledge will be self-evident. Currently, the usefulness of standards is not likely to be realized because there are differences in meaning and methods between disciplines. We would suggest that metadata needs to provide information about the *usefulness* of the data rather than its *usability*. However, as any measure or qualification of usefulness can only be relative to the intended application, new forms of metadata must be suggested. Schuurman and Leszczynski (2006) and Comber *et al.* (2005b) have argued that metadata ought to include semantics and report more than documentation of the technical aspects of data production. This raises two parallel and perhaps mutually incompatible arguments, either of which could be used to provide the concluding paragraph to this commentary:

1)      the current situation for users of spatial metadata is analogous to buying a car just on the basis that it has a certificate of road worthiness. Standards have been created that state each car must have a safety certificate, and car manufacturers ensure that their cars conform to those standards. A mechanical test checks that the cars conform to those standards each year. Mediators (car dealers) use the safety certificate to show that the car is of merchantable quality. The user (buyer) is reassured that the car is safe and conforms to the law. However, in reality the purchaser has to consider whether the car conforms to their current (and projected future) needs and circumstances. This process to identify the *usefulness* of different vehicles relative to their needs may include taking a test drive, using independent assessment (a consumer report, motoring magazines, experiences of friends and neighbours, etc.). At the moment there are no such tools or independent assessments available to users of spatial data. This is partly because, unlike the producers of cars, the producers of spatial data have been monopoly suppliers, but for some that is changing.

2)      the implications of the arguments made in this paper suggest that metadata that includes semantics might stand-in as a training in the specific discipline of the data. This presents some interesting metadata conundrums. Soil orders reflect the view that the history of the soil may be important, which mirrors the emphasis placed on evolutionary history in plant and animal taxonomies. However the evolution of soils is interesting rather than important to users and metadata. Information that is important in disciplinary and non-disciplinary contexts will always exist. So what levels of metadata are appropriate? The alternative concluding statement could be to argue that "any understanding provided by metadata would be insufficient for effective use of spatial information which depends on real understanding of what the data means. This implies specialist Masters-level training and questions the need and purpose of metadata for anything other than discovery."

We hope that the issues raised in this paper will stimulate discussion on the role and content of metadata for geographical information.

## Acknowledgments

Some of the ideas in this paper were first presented in Vienna at *Spatial Data Handling 2006* and originate from work within the REVIGIS project funded by the European Commission (Project Number IST-1999-14189) and during a workshop sponsored by the National Institute for Environmental eScience in 2005. We thank Rodolphe Devillers and the Programme Committee of *ISSDQ 2009* for their invitation to present our ideas as a keynote paper.

# References

ANZLIC (2001), *Metadata Guidance: core metadata elements for geographic information in Australia and New Zealand*, Griffith ACT, Australia, 95p.

Brummitt, R.K. (2001), *World Geographical Scheme for Recording Plant Distributions Edition 2, Plant Taxonomic Database Standards No 2*, Hunt Institute for Botanical Documentation, Pittsburgh, 137 p.

Calvani, D. (2004), *Between interpreting and cultures: a community interpreters toolkit* http://www.aucegypt.edu/ResearchatAUC/rc/fmrs/reports/Documents/Annual_Report200 4_Edited_Final.pdf, accessed 12 February 2007.

Cherrill, A. and C. McClean. (1999), "Between-observer variation in the application of a standard method of habitat mapping by environmental consultants in the UK". *Journal of Applied Ecology*, Vol. 36: 989-1008.

Comber, A.J., P.F. Fisher and R.A. Wadsworth. (2003), "Actor Network Theory: a suitable framework to understand how land cover mapping projects develop?" *Land Use Policy*, Vol. 20: 299-309.

Comber, A.J., P.F. Fisher and R. Wadsworth. (2004), "Integrating land cover data with different ontologies: identifying change from inconsistency". *International Journal of Geographical Information Science*, Vol. 18: 691-708.

Comber, A.J., P.F. Fisher and R.A. Wadsworth. (2005a), "What is land cover?" *Environment and Planning B: Planning and Design*, Vol. 32: 199-209.

Comber, A.J., P.F. Fisher and R.A. Wadsworth. (2005b), "You know what land cover is but does anyone else?...an investigation into semantic and ontological confusion". *International Journal of Remote Sensing*, Vol. 26: 223-228.

Comber, A.J., P.F. Fisher, F. Harvey, M. Gahegan and R.A. Wadsworth. (2006), "Using metadata to link uncertainty and data quality assessments". *In:* A. Riedl, W. Kainz, and G. Elmes (eds.). *Progress in Spatial Data Handling, Proceedings of SDH 2006*, Springer, Berlin, pp. 279-292.

Comber, A.J., P.F. Fisher and R.A. Wadsworth. (2008), "Semantics, Metadata, Geographical Information and Users". *Transactions in GIS*, Vol. 12: 287-291.

Cowardin, L.M., V. Carter, F.C. Golet and E.T. LaRoe. (1979), *Classification of Wetlands and Deepwater Habitats of the United States National Wetland Inventory, US Department of the Interior*, Washington, DC Adopted as FGDC-STD-004 http://el.erdc.usace.army.mil/emrrp/emris/emrishelp2/cowardin_report.htm, accessed 6 April 2009.

Darwin Core (2008), *Purpose and design goals of the Darwin Core*, http://wiki.tdwg.org/twiki/bin/view/DarwinCore/DesignAndPurpose, accessed 2 April 2009.

DCDSTF (Digital Cartographic Data Standards Task Force) (1988), "The proposed standard for digital cartographic data". *American Cartographer*, Vol. 15: 9-140.

Di Gregorio, A. and L.J.M. Jansen. (2000), *Land Cover Classification Scheme (LCCS): Classification Concepts and User Manuals*, Version 1.0, Food and Agricultural Organisation of the United Nations, Rome, 179p.

Eason, B., J. Gregory, B. Drach, K. Taylor and S. Hanki. (2009), *NetCDF Climate and Forecast (CF) Metadata Convention Version 1.4*, http://cf-pcmdi.llnl.gov/documents/cf-conventions/1.4/cf-conventions.pdf, accessed 6 April 2009.

FGDC (Federal Geographic Data Committee) (1997a), *Vegetation Classification Standard FGDC-STD-005*, Vegetation Subcommittee, Reston, Virginia http://www.fgdc.gov/standards/projects/FGDC-standards-projects/vegetation/vegclass.pdf, accessed 2 April 2009.

FGDC (Federal Geographic Data Committee) (1997b), *Soil Geographic Data Standards FGDC-STD-006*, Soil Data Subcommittee, National Technical Information Service, Computer Products Office, Springfield, Virginia, USA. http://www.fgdc.gov/standards/projects/FGDC-standards-projects/soils/soil997.PDF, accessed 2 April 2009.

FGDC (Federal Geographic Data Committee) (1998), *Content Standard for Digital Geospatial Metadata, FGDC-STD-001-1998*, Reston, Virginia. http://www.fgdc.gov/standards/projects/FGDC-standards-projects/metadata/base-metadata/v2_0698.pdf, accessed 2 April 2009.

FGDC (Federal Geographic Data Committee) (1999), *Content Standard for Digital Geospatial Metadata, Part 1: Biological Data Profile FGDC-STD-001.1-1999*, Biological Data Working Group, Reston, Virginia. http://www.fgdc.gov/standards/projects/FGDC-standards-projects/metadata/biometadata/biodatap.pdf, accessed 2 April 2009.

FGDC (Federal Geographic Data Committee) (2001), *Shoreline Metadata Profile of the Content Standards for Digital Geospatial Metadata FGDC-STD-001.2-2001*, Marine and Coastal Spatial Data Subcommittee, Reston, Virginia. http://www.csc.noaa.gov/metadata/sprofile.pdf, accessed 2 April 2009.

FGDC (Federal Geographic Data Committee) (2003), *Cadastral Data Content Standard for the National Spatial Data Infrastructure, FGDC-STD-003 Version 1.3*, Subcommittee on Cadastral Data, Reston, Virginia. http://www.nationalcad.org/data/documents/CADSTAND.v.1.3.pdf, accessed 6 April 2009.

Fisher, P.F. (1998), "Improved Modeling of Elevation Error with Geostatistics". *GeoInformatica*, Vol. 2: 215-233.

Fisher, P.F. (2003), "Multimedia Reporting of the Results of Natural Resource Surveys". *Transactions in GIS*, Vol. 7: 309-324.

Goodchild, M.F. (2006), "GIScience Ten Years After Ground Truth". *Transactions in GIS*, Vol. 10: 687-692.

Harvey, F. and N. Chrisman. (1998), "Boundary objects and the social construction of GIS technology". *Environment and Planning A*, Vol. 30: 1683-1694.

Herold, M., J.S. Latham, A. Di Gregorio and C.C. Schmullius. (2006), "Evolving standards in land cover characterization". *Journal of Land Use Science*, Vol. 1: 157-168.

Hunter, G.J. (2001), "Spatial Data Quality Revisited". *In: Proceedings of GeoInfo 2001*, 4-5 October, Rio de Janeiro, Brazil, pp. 1-7.

INSPIRE (Infrastructure for Spatial Information in Europe) (2007), Draft Guidelines – INSPIRE metadata implementing rules based on ISO 19115 and ISO 19119. *http://inspire.brgm.fr/Documents/MD_IR_and_ISO_20080425.pdf*, accessed 6 April 2009.

ISO (2001), *ISO 19110 Geographical Information – Methodology for feature cataloguing*, International Standards Organisation, Geneva, 53p.

ISO (2003a), *ISO 19115 Geographical Information – Metadata*, International Standards Organisation, Geneva, 140p.

ISO (2003b), *ISO 19114 Geographical Information – Quality Evaluation Procedures*, International Standards Organisation, Geneva, 63p.

Kresse, W. and K. Fadaie. (2004), *ISO Standards for Geographic Information*. Springer, Berlin, 322p.

Lawrence, B.N., R. Cramer, M. Gutierrez, K. van Dam, S. Kondapalli, S. Latham, R. Lowry, K. O'Neill and A. Woolf. (2003), "The NERC DataGrid Prototype". *In:* S.J. Cox and D.W. Walker (eds.) *Proceedings of the UK e-Science All Hands Meeting, April 2003*, http://ndg.nerc.ac.uk/public_docs/AHM-2003-BNL.pdf, accessed 6 April 2009.

Martin, E. (2000), "Actor–networks and implementation: examples from conservation GIS in Ecuador". *International Journal of Geographical Information Science,* Vol. 14: 715-737.

Merriam-Webster (2009), *Merriam-Webster Online,* http://www.merriam-webster.com/dictionary/semantics, accessed 2 April 2009.

Moellering, M. (ed.) (2005), *World Spatial Metadata Standards: Scientific and Technical Characteristics, and Full Descriptions with Crosstable.* International Cartographic Association, Oxford, Pergamon, 710p.

OGC (Open Geospatial Consortium) (2007), *Observations and Measurements – Part 1 - Observation schema.* http://www.opengeospatial.org/standards/om accessed 2 April 2009.

Ordnance Survey (2009), "OS MasterMap Topography Layer".
http://www.ordnancesurvey.co.uk/oswebsite/products/osmastermap/layers/topography/index.html, accessed 2 April 2009.

Salgé, F. (1995), "Semantic Accuracy". *In:* S.C. Guptill and J.L. Morrison (eds.), *Elements of Spatial Data Quality,* Elsevier, Oxford, pp. 139-151.

Schuurman, N. and A. Leszczynski. (2006), "Ontology-Based Metadata". *Transactions in GIS,* Vol. 10: 709-726.

Shortridge, A., J. Messina, S. Hession and Y. Makido. (2006), "Towards an Ontologically-driven GIS to Characterize Spatial Data Uncertainty". *In:* A. Riedl, W. Kainz, and G. Elmes (eds.), *Progress in Spatial Data Handling, Proceedings of SDH 2006,* Springer, Berlin, pp. 465-476.

Soil Survey Staff (1999), *Soil Taxonomy, A Basic System of Soil Classification for Making and Interpreting Soil Surveys 2nd Edition.* Natural Resources Conservation Service Number 436, US Government Printing Office, Washington, DC 20402.

Van de Vlag, D., B. Vasseur, A. Stein and R. Jeansoulin. (2005), "An application of problem and product ontologies for the revision of beach nourishments". *International Journal of Geographical Information Science,* Vol. 19: 1057-1072.

Visser, U., H. Stuckenschmidt, G. Schuster and T. Vogele. (2002), "Ontologies for geographic information processing". *Computers and Geosciences,* Vol. 28: 103-117.

Wadsworth, R.A., A.J. Comber and P.F. Fisher. (2006), "Expert knowledge and embedded knowledge: or why long rambling class descriptions are useful". *In:* A. Riedl, W. Kainz and G. Elmes (eds.). *Progress in Spatial Data Handling, Proceedings of SDH 2006,* Springer, Berlin, pp. 197-213.

# 2

## Scale Is Introduced in Spatial Datasets by Observation Processes

*Andrew U. Frank*

### Abstract

*An ontological investigation of data quality reveals that the quality of the data must be the result of the observation processes and the imperfections of these. Real observation processes are always imperfect. The imperfections are caused by (1) random perturbations, and (2) the physical size of the sensor. Random effects are well-known and typically included in data quality descriptions. The effects of the physical size of the sensor limit the detail observable and introduce a scale to the observations. The traditional description of maps by scale took such scale effects into account, and must be carried forward to the data quality description of modern digital geographic data. If a sensor system is well-balanced, the random perturbations, size of the sensor and optical blur (if present) are of the same order of magnitude and a summary of data quality as a 'scale' of a digital data set is therefore theoretically justifiable.*

### 2.1  Introduction

Digital geographic data comes in different qualities, and applications have different requirements for the quality of their inputs. A common misconception is that better quality is always preferable, forgetting that better quality means more detail and therefore more data, longer data transfer and processing time, etc. Traditionally *map scale* was used to describe the quality of geographic data comprehensively. With the reduction in scale, expressed as representative fraction, comes automatically a reduction in detail, described as cartographic generalization. Users learned which map scales were suitable for which task: orienteering uses maps in the scale range 1:10,000 to 1:25,000; for driving from city to city, maps 1:250,000 to 1:500,000 are sufficient, etc. Repeated experiences have taught us these practical guidelines and we follow them without asking for an underlying theory. The beginning of such a theory is attempted here.

In the age of digital data, the traditional definition of scales, as proportion between distances on the map and in reality, does not make sense: locations are expressed with coordinates and distances computed are in real world units. Only when preparing a graphical display is a numeric scale used – but, in principle, digital geographic data can be shown at any desired graphical scale, even if this often does lead to nonsense. The concept of scale in a digital world has been critically commented, but no solution suggested (Lam and Quattrochi, 1992; Goodchild and Proctor, 1997; Reitsma and Bittner, 2003). The discussion of data quality of geographic data focuses on descriptions of data quality of a single dataset, sometimes differentiating different types of digital representation (e.g., Goodchild, 1994). I want to complete this dataset viewpoint with an analysis of the process by which data is produced from observations and used in discussions about actions.

In this paper I explore the process of geographic data collection and show how scale is introduced during the observation process, and should be carried forward as a quality indication. An analysis of the properties of real (physical) observation processes reveals that physical observation processes introduce a scale into the observation. This 'scale' is not an artifact of cartography, but originates in the physical observation process itself. The same 'scale' value can later be used when considering whether a dataset can be used effectively in a decision situation (Frank, 2008).

This paper begins with a short review of tiered ontology (Frank, 2001; Frank, 2003), which is necessary for the analysis. Section 2.3 lists the information processes that link the tiers and gives the framework used. Section 2.4 discusses briefly accuracy and shows how random imperfections in the observations influence the formation of objects and the values for their attributes. Section 2.5 looks at scale, produced by the spatial and temporal extent of the observation as a second source of imperfection in the data. Convolution gives a formal model for this effect. The influences of scale are so defined in the process from observation to object related data.

Goodchild (1994) points to the spatial extent necessarily associated with some types of geographic data; for example "land cover" (Goodchild, 1994: 616). Unfortunately, his contribution had more impact on vegetation mapping than on spatial data quality research. The present contribution extends Goodchild's observation and states that primary observations have necessarily a scale and that the scale of derived datasets can be traced through the information processing steps. In particular, the novel contribution of this paper is: (1) identifying that the scale of data is introduced during the observation process; (2) providing a formal model that can be used to predict effects of the scale of observations on other data.

## 2.2 Tiered Ontology

An ontology describes the conceptualization of the world used in a particular context (Guarino and Poli, 1995; Gruber, 2005): two different applications use generally different conceptualizations. The ontology clarifies these concepts and communicates the semantics intended by data collectors and data managers to persons making decisions with the data. An ontology for an information system that separates different aspects of reality must not only conceptualize the objects and processes in reality, but must also describe the information processes that link the different conceptualizations and transform between them. This is of particular importance for an ontology that divides conceptualization of reality in tiers (Frank, 2001) or in object and process ontologies (Bittner and Smith, 2003; Smith and Grenon, 2004). The processes transform the data and the quality of the data; understanding and formalizing the information processes allow one to describe how the quality of the observation determines the quality of derived data.

The tiered ontology used here (Frank, 2001; Frank, 2003) starts with tier O, which is the physical reality, the real world, that "what is?", independent of human interaction with it. Tier O is the Ontology proper in the philosophical sense; sometimes Ontology in this sense is capitalized and it is never used in a plural form. In contrast, the ontologies for information systems, which are the focus of this paper, are written with a lower case o.

### 2.2.1  Tier 1: Point observations

Reality is observable by humans and other cognitive agents (e.g., robots, animals). Physical observation mechanisms produce data values from the properties found at *a point in space and time*; $v=p(x, t)$. The value $v$ is the result of an observation process $p$ of physical reality found at point $x$ and time $t$.

Tier 1 consists of the data resulting from observations at specific locations and times (termed *point observation*). Examples would be the temperature, type of material, or forces at a point. In GIS such observations are often realized as raster data resulting from remote sensing (Tomlin, 1983), similar to the observations collected by our retina, which performs many observations of light emanating from a point in parallel. Sensors, and sensor networks in general, also produce point observations. Many, if not most measurements performed in the world are more complex, and report attributes of objects (e.g., length, weight) and are part of tier 2. Point observations are so simple that they are assumed as functions, unlike complex measurements of object properties influenced by culture and conventions (e.g., published standards or regulations).

### 2.2.2  Tier 2: Objects

The second tier is a mental description of the world in terms of mentally constructed physical objects. Objects are regions of space that have some uniformity in property. An object representation reduces the amount of data, if the subdivision of the world into objects is such that most properties of the objects remain invariant in time (McCarthy and Hayes, 1969). For example, most properties of a taxi cab remain the same for hours, and days, and need not be observed and processed repeatedly; only location and occupancy of the taxi cab change often. The critical question is how mental objects are constructed, subjectively and in response to concrete situations.

### 2.2.3  Tier 3: Social constructions

Tier 3 consists of constructs combining and relating physical objects to abstract constructs. These are first conventions to allow communication, which link mental objects (thoughts) with symbols, and second constructions like money, marriage and land ownership. Constructed reality links a physical object X to mean the constructed object Y in the context Z. The formula is: *"X counts as Y in context Z"* (Searle, 1995: 28).

Social constructions relate physical objects or processes to abstract constructs of objects or process type connected by human perception and mental object formation shaped by cultural conventions. Constructed objects can alternatively be constructed from other constructed objects, but all constructed objects are eventually grounded in physical objects.

## 2.3  Information Processes

Information processes transform information obtained at a lower tier to a higher tier (Figure 2.1).

Ontological Tier        Information
                        Process

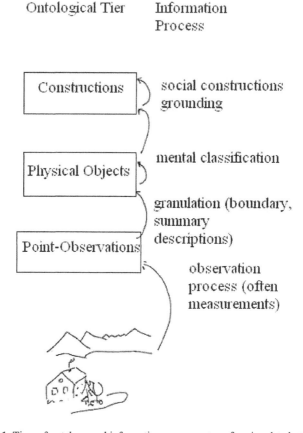

**Figure 2.1.** Tiers of ontology and information processes transforming data between them.

All human knowledge is directly or indirectly the result of point observations, transformed in a chain of information processes to complex knowledge about mentally constructed objects. *All imperfections in data must be the result of some aspect of an information process.* As a consequence, all theory of data quality and error modeling has to be related to empirically justified properties of the information processes.

### 2.3.1 Observations of physical properties at point

The observations of physical properties at a specific point are the *physical process* that links tier O to tier 1; the realization of observations is unavoidably imperfect in two ways:

1. unpredictable random disturbance in the value produced, and
2. observations focus not at a point but over an extended area.

A systematic bias, if present, can be included in the model of the sensor and be corrected by a function and is not considered further.

### 2.3.2  Object formation (granulation)

The formation of objects – what Zadeh calls *granulation* (Zadeh, 2002) – is a complex process of determining, first, the boundaries of objects and, second, summarizing properties for the delimited regions. Gibson (1986) posits that humans create a *meaningful environment* mentally consisting of *meaningful things*, which I call (mental) objects. For objects on a table top (Figure 2.2), e.g., a coffee cup, a single process of object formation dominates: we form spatially cohesive solids, which move as a single piece: a cup, a saucer, and a spoon.

**Figure 2.2.** Objects on a tabletop.

Geographic space does not lend itself to such a single, dominant, subdivision as objects typically do not move. Various aspects can be used to form regions of uniform properties, leading to different objects overlapping in the same space. Watersheds, areas above a particular height, regions of uniform soil, uniform land management (Figure 2.3), and so on, can all be meaningful objects in a situation (Couclelis and Gottsegen, 1997). Object classification forms groups of objects suitable for certain operations (hunting, planting crops, grazing cattle, etc.). Processes can be granulated by similar approaches in 3D plus time (Reitsma and Bittner, 2003), an important future research topic.

**Figure 2.3.** Fields on a mountain.

We are not aware that our eyes (but also other sensors in and at the surface of our body) report point observations. The individual sensors in the eye's retina give a pixel-like observation, but the eye seems to report size, color, and location of objects around us. The observations are, immediately and without the person being conscious about the processes involved, converted to object data, and mental processes of object formation connect tier 1 to tier 2. Most of the technical systems we use to measure object attributes hide in a similar way the intricacies of the object formation: a balance reports the total weight of the 'object', i.e., all that is placed in the weighing pane. Length measure report about comparison of length between an object of interest and a yardstick. Scheider and Kuhn (2008) describe similar, but virtual (imagined), operations related to the linguist's fictive motion (Talmy, 1996).

Object formation increases the imperfection of data – instead of having detailed knowledge about each individual pixel, only a summary description is retained. This summary may average out some of the imperfections of the point observations and the result may be more useful. Reporting information with respect to objects results in a substantial reduction in size of the data (estimations for the amount of data to represent, for example, a field in figure 2.3 as raster or vector suggest a data compression factor as high as 106).

Object formation consists itself of two information processes, namely, (1) boundary identification, and (2) computing summary descriptions.

### 2.3.2.1  Boundary identification

Objects are formed as regions in 2D or 3D that are uniform in some aspect. The dominant approach is to identify surfaces and define objects as "things which move in one piece". This uniformity in marginal coherence works for objects in tabletop space (Figure 2.2), but fails for geographic space because there is not one dominant way to partition the real world into objects, but several, depending on the viewpoint and situation. In order to form objects, a property that is important for the current situation is selected to be uniform. A rural field is uniform in its land cover; tabletop objects are uniform in material coherence and in their movement. Note that object formation exploits the strong correlation found in the real world; human life would not be possible in a world without strong spatial and temporal correlation. The details of how objects are identified are determined by the interactions intended with them.

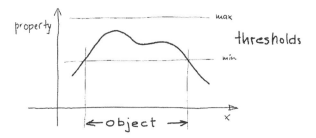

**Figure 2.4.** The property, which should be uniform within some threshold values determines the object boundaries.

An object boundary is determined by first selecting a property and a property value that should be uniform across the object, similar to the well-known procedure for regionalization of 2D images. The boundary is at a threshold for this value, or at the place where the property changes most rapidly (Figure 2.4).

### 2.3.2.2 Determination of descriptive summary data (attributes of objects)

Descriptive values summarize the properties of space within the object limited by a boundary. The computation is typically a function that determines the sum, maximum, minimum, or average over the region, for example, total weight of a movable object, amount of rainfall on a watershed, maximum elevation in a country (Tomlin, 1983; Egenhofer and Frank, 1986).

### 2.3.3  Classification

Objects, once identified, are classified. On the tabletop we see cups, spoon, and saucers; in a landscape, forest, fields, and lakes are identified. Mental classification relates the objects identified by granulation processes to operations, i.e., interactions of the cognitive agent with the world. Such actions are comparable to Gibson's affordances (Gibson, 1986; Raubal, 2002) when performing an action. To illustrate, to pour water from a pitcher into a glass requires a number of properties of the objects involved: the pitcher and the glass must be containers, i.e., having the affordance to contain a liquid, the object poured must be a liquid, and so on.

Potential interactions between the agent and objects, or interactions of interest between objects, assert conditions these objects must fulfill, expressed as an attribute and a range for the value of the attribute. I have used the term *distinction* for the differentiation between objects that fulfill a condition and those that do not. Distinctions are partially ordered: a distinction can be finer than another one (e.g., 'drinkable' is a subtaxon of 'liquid'), and distinctions form a taxonomic lattice (Frank, 2006).

### 2.3.4  Constructions

Tier 3 contains constructions, which are linked through granulation and mental classification to the physical objects and operations. They are directly dependent on the information processes described above, but details are not consequential for present purposes.

## 2.4  Random Effects in the Observations

A physical sensor cannot realize a perfect observation at a point in space and time. Physical sensors are influenced by random processes that produce perturbations of the observations. The unpredictable disturbance is typically modeled by a probability distribution. For most sensors a normal (Gaussian) probability distribution function (PDF) is an appropriate choice. This model is, after correction for systematic influences and bias, applicable to a very wide variety of point observations.

### 2.4.1  Influence on object formation

Errors in observation influence the determination of the object boundary. The statistical error of the boundary for simple cases follows from Gauss' Law of error propagation (Frank, 2008) (Figure 2.5 does not include an uncertainty in the thresholds for graphical clarity; though the influence would be similar). The summary values are similarly influenced by random perturbations in the observations (it is likely that random effects are reduced by averaging).

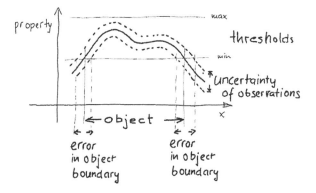

**Figure 2.5.** The influence of uncertainty in the observations creates an uncertainty in the object boundary.

If the observation information processes allow a probabilistic description of the imperfections of the values, then the imperfections in the object boundary and summary value are equally describable by a probability distribution. Assuming a PDF for the property used for the determination of the boundary, one can describe the PDF for the boundary line. It is an interesting question whether the PDF transformation functions associated with boundary derivation and derivation of summary values preserve a normal distribution, i.e., if the observation processes described by imperfections with a normal distribution produce imperfections in boundary location and summary values, which are again describable by a normal distribution. Further studies may show that the effects are multiplicative and produce a Rayleigh-like distribution, or that imperfections of the processes are correlated, posing the difficult problem of estimating the correlations.

### 2.4.2  Classifications

Distinctions reflect the limits in the property values of an object, where the object can or cannot be used for a specific interaction. The decision whether or not the values for an object are inside the limits is more or less sharp, and the cutoff usually gradual. Class membership is therefore a fuzzy membership functions as originally defined by Zadeh (1974).

## 2.5  Scale in Observations

In this section the effects of the finiteness of physical observation instruments are discussed. Physical observation instruments may be very small, but not infinitely small. A sensor cannot realize a perfect observation at a perfect point in space or time. Any physical observation integrates a physical process over a region during a time. The time and region over which the integration is performed can be very small (e.g., a pixel sensor in a camera has a size of 5/1000mm and integrates the photons arriving in this region for as little as 1/5000 sec) but it is always of finite size and duration. The size of the area and the duration influence the result.

The effects of the size of the observation device is as equally unavoidable as the random perturbations of the observations, which is more widely recognized, discussed, and given a formal model (summarized in the previous section). This section proposes a formal model for the finiteness of the observation device. The sensor can be modeled as a convolution with a Gaussian kernel applied to the function representing the physical reality.

The necessary finiteness of the sensor introduces an unavoidable scale element in the observations. Scale effects are not yet well understood, despite many years of being listed as one of the most important research problems (Abler, 1987; NCGIA, 1989a; NCGIA, 1989b; Goodchild *et al.*, 1999), and it is expected that the formalization given here advances this research.

### 2.5.1  Physical point-like observation

The intended point-observation $v = f(x, t)$ cannot be realized, but the observation device reports the average value for a small area and a small time interval,

$$v(x,t) = \int_{-\varepsilon}^{+\varepsilon} f((x,t) - \varepsilon) d\varepsilon$$

where the multidimensional space-time vector value $\varepsilon$ ranges over the size and the time interval used by the sensor (corresponding to the values of $x$ and $t$). Convolution with a kernel $k(\varepsilon)$ is a formal model for the real observation at point,

$$v(x,t) = \int f((x,t) - \varepsilon) k(\varepsilon) d\varepsilon$$

The value $f(x, t)$ is multiplied by the kernel value $k(\varepsilon)$, which is non-zero for a small region only around zero and for which,

$$\int k(\varepsilon(\varepsilon)) = 1$$

For sensors in cameras, a rectangular kernel or a Gaussian Kernel is assumed; the latter is optimal to satisfy the sampling theorem. Convolution with a Gaussian kernel produces an average effect on the signal.

### 2.5.2 Sampling theorem

The sampling theorem addresses another related limitation of real observations: it is impossible to observe infinitely many points; real observations are limited to *sampling* the phenomenon of interest at finitely many points. Sampling introduces the danger that the observations include spurious signals not present in reality (aliasing). The Nyquist-Shannon *sampling theorem* states that sampling must be twice as frequent as the highest frequency in the signal to avoid artifacts. If the sampling rate is fixed, the signal must be filtered and all frequencies higher than half the sampling frequency cut off (low-pass filter). In the audio world the sampling theorem is well-known, but it applies to any dimension, including sampling by remote sensing or sensor network in geographical space. The sampling theorem applies to remote sensing and sensors are appropriately designed, but it is less discussed in geography and geographical information science. It may appear strange to speak of spatial frequency, but it is effective to make the theory available to GIScience, where it applies to all dimensions observed (2 or 3 spatial dimensions, temporal, etc.).

### 2.5.3 Scale of observations

Observations modeled as convolution with a Gaussian Kernel are effectively applying a suitable low-pass filter, and produces valid results. The size of the non-zero region of the kernel $k(\varepsilon)$ affects the observation result, but is not part of the physical reality observed, instead being caused by the observation system. The observations are influenced by the size of the non-zero region of the kernel. For this situation it appears reasonable to say that the observation has the 'scale' corresponding to $2\ \sigma$ of the kernel and half the sampling frequency $v$. A numerical description of 'scale' could use the value for $\sigma$ with dimension time$^{-1}$ (second$^{-1}$) and length$^{-1}$ (meter$^{-1}$) respectively. If a unit of 1mm$^{-1}$ is selected, then the numerical value is comparable to traditional map scale denominators but not dimensionless.

It is debatable whether to call this scale, adding one more sense to the close to a dozen already existing (Wikipedia lists eleven), or to use a term like *resolution* or *granularity*. I prefer 'scale' because speaking of a dataset and stating its 'scale', for example as "30,000 mm$^{-1}$ in space and 5 years$^{-1}$ in time", extends the current usage reasonably and describes typical topographic maps.

### 2.5.4 Effects of scales on object formation

**Size of smallest objects detected**: the scale of the observation limits the smallest object that can be detected; objects with one less dimension than the scale are not observed, and their extent is aggregated with the neighbors. This is comparable to the cartographic minimal mapping unit. In data of a scale $m$ one does not expect objects smaller than $f \cdot m^2$, where $f$ is a form descriptor indicating how different an object is from a square or circle (respective cube or sphere).

**Effects on uniformity**: differences in property values less than the scale (for this property) are not observable, and are therefore not available when differentiating two objects; this is essentially the effect described above that small objects escape detection; the small separating object is not being observable.

**Effects on attribute values**: if attribute values are desired as summary values over the area of the object, the effects of the scales of observation in the property used to derive the attribute will statistically cancel – if the scale of the observation used for object formation and the property integrated is comparable. Averages (results of integrals) tend to be less extreme the larger the scale as a result of averaging in the observation process.

**Effects on object classification**: the scale of observation influences directly the object formation and indirectly the classification. This is most important if the class is distinct by size, e.g., small buildings vs. larger buildings.

## 2.6 Conclusions

Physical observation systems deviate in two inevitable and non-avoidable respects from the perfect point observation of the properties of reality: *random* perturbation of results, and *finite* spatial and temporal *extent* over which the observation system integrates.

Using a tiered ontology where point observations are separated from object descriptions allows one to follow how the imperfection introduced by random error and scale propagate to objects and their attributes. It was shown that precision and scale are valid descriptors of datasets, and originate with unavoidable imperfections of physical observations. Random effects are described by a probability distribution function (PDF), and the propagation of effects of random errors follow, in simple cases, Gauss' Law of error propagation; in general a transformation for the PDF is computed.

The effects of finite support for the observation can be modeled as a convolution with a Gaussian Kernel, and the non-zero extent of the kernel determines the 'scale' of the observation. The signal must be filtered with a low-pass filter to cut off all frequencies above half the sampling frequency. Convolution with a Gaussian kernel achieves this approximately.

The description here used a prototypical remote sensing observation as sensor, which produces point observations where the low-pass filter is implied in the observation system. It is recommended to investigate how the sampling theorem applies to sensor networks and other geographic observations. Well-designed observation systems are balanced such that effects of (1) random perturbation, (2) the extent of the sensors, and (3) the blur of the optical system (if any) produce imperfections of comparable size. Datasets produced by well-designed, balanced observation systems can be characterized by scales – as was traditionally done.

Geographic Information Systems are used to combine data from different sources; the theory outlined above shows how to treat cases where not all data have the same scale. In particular, it was shown how the notion of scale applies to so-called vector (object) data sets (Goodchild, 1994) and traces it back to the known methods for raster data.

## Acknowledgements

These ideas were developed systematically for a talk I presented at the University of Münster. I am grateful to Werner Kuhn and his colleagues for providing me with this opportunity. Three anonymous reviewers provided very valuable feedback, and helped extend and clarify the arguments in the paper.

## References

Abler, R. (1987), "Review of the Federal Research Agenda". *In:* R.T. Aangenbrug *et al.* (eds) *International Geographic Information Systems (IGIS) Symposium (IGIS'87): The Research Agenda, Arlington, VA*, NASA.

Bittner, T. and B. Smith. (2003), "Granular Spatio-Temporal Ontologies". *In:* H.W. Guesgen, D. Mitra and J. Renz (eds) *2003 AAAI Symposium: Foundations and Applications of Spatio-Temporal Reasoning (FASTR)*, AAAI Press, Menlo Park Cal: 6.

Couclelis, H. and J. Gottsegen. (1997), "What Maps Mean to People: Denotation, Connotation, and Geographic Visualization in Land-Use Debates". *In:* S.C. Hirtle and A.U. Frank (eds.) *Spatial Information Theory - A Theoretical Basis for GIS (International Conference COSIT'97)*. Berlin-Heidelberg, Springer-Verlag. Lecture Notes in Computer Science, Vol. 1329: pp. 151-162.

Egenhofer, M. and A.U. Frank. (1986), "Connection between Local and Regional: Additional "Intelligence" Needed". *FIG XVIII International Congress of Surveyors, Toronto, Canada (June 1-11, 1986)*.

Frank, A.U. (2001), "Tiers of Ontology and Consistency Constraints in Geographic Information Systems". *International Journal of Geographical Information Science*, Vol. 75(5 (Special Issue on Ontology of Geographic Information)): 667-678.

Frank, A.U. (2003), "Ontology for Spatio-Temporal Databases". *In:* M. Koubarakis, T. Sellis, A.U. Frank, S. Grumbach, R.H. Güting, C.S. Jensen, N. Lorentzos, Y. Manopoulos, E. Nardelli, B. Pernici, H.-J. Schek, M. Scholl, B. Theodoulidis and N. Tryfona (eds.) *Spatiotemporal Databases: The Chorochronos Approach*. Berlin, Springer-Verlag: pp. 9-78.

Frank, A.U. (2006), "Distinctions Produce a Taxonomic Lattice: Are These the Units of Mentalese?" *International Conference on Formal Ontology in Information Systems (FOIS)*, IOS Press, Baltimore, Maryland, pp. 27-38.

Frank, A.U. (2008), "Analysis of Dependence of Decision Quality on Data Quality." *Journal of Geographical Systems*, Vol. 10(1): 71-88.

Gibson, J.J. (1986), *The Ecological Approach to Visual Perception*, Lawrence Erlbaum, Hillsdale, NJ, 332p.

Goodchild, M.F. (1994). "Integrating GIS and Remote Sensing for Vegetation Analysis and Modeling: Methodological Issues". *Journal of Vegetation Science*, Vol. 5: 615-626.

Goodchild, M.F., M.J. Egenhofer, K.K. Kemp, D.M. Mark and E. Sheppard. (1999), "Introduction to the Varenius Project". *International Journal of Geographical Information Science*, Vol. 13(8): 731-745.

Goodchild, M.F. and J. Proctor. (1997), "Scale in a Digital Geographic World". *Geographical & Environmental Modelling*, Vol. 1(1): 5-23.

Gruber, T. (2005), "TagOntology – a way to agree on the semantics of tagging data". http://tomgruber.org/writing/index.htm (accessed April 22, 2009).

Guarino, N. and R. Poli (eds) (1995), "Formal Ontology, Conceptual Analysis and Knowledge Representation". Special issue of the *International Journal of Human and Computer Studies*, Vol. 43(5/6).

Lam, N. and D.A. Quattrochi. (1992), "On the issues of scale, resolution, and fractal analysis in the mapping sciences". *The Professional Geographer*, Vol. 44:88-98.

McCarthy, J. and P.J. Hayes. (1969), "Some Philosophical Problems from the Standpoint of Artificial Intelligence". *In:* B. Meltzer and D. Michie (eds.) *Machine Intelligence 4*, Edinburgh University Press, Edinburgh, pp. 463-502.

NCGIA. (1989a), "The U.S. National Center for Geographic Information and Analysis: An Overview of the Agenda for Research and Education." *International Journal of Geographical Information Science*, Vol. 2(3): 117-136.

NCGIA. (1989b), *Use and Value of Geographic Information. Initiative Four Specialist Meeting, Report and Proceedings*, National Center for Geographic Information and Analysis; Department of Surveying Engineering, University of Maine; Department of Geography, SUNY at Buffalo.

Raubal, M. (2002), *Wayfinding in Built Environments: The Case of Airports*. (IfGIprints Vol. 14). Institut für Geoinformatik, Institut für Geoinformation, Münster, Solingen.

Reitsma, F. and T. Bittner. (2003), "Scale in Object and process Ontologies". *In:* W. Kuhn, M.F. Worboys and S. Timpf (eds) *Foundations of Geographic Information Science, International Conference, COSIT 2003, Ittingen, Switzerland, September 24-28, 2003, Proceedings*, Springer-Verlag, Berlin, pp. 13-27.

Scheider, S. and W. Kuhn. (2008), "Road Networks and Their Incomplete Representation by Network Data Models". *Proceedings of the 5th International Conference on Geographic Information Science* (Lecture Notes in Computer Science, Vol. 5266). Springer-Verlag, Berlin, pp. 290-307.

Searle, J.R. (1995), *The Construction of Social Reality*. New York, The Free Press, 241p.

Smith, B. and P. Grenon. (2004), "SNAP and SPAN: Towards Dynamic Spatial Ontology". *Spatial Cognition and Computing*, Vol. 4: 69-103.

Talmy, L. (1996), "Fictive Motion in Language and "Ception"". *In:* P. Bloom, M.A. Peterson, L. Nadel and M.F. Garett (eds.) *Language and Space*, MIT Press, Cambridge, MA, pp. 211-276.

Tomlin, C.D. (1983), "A Map Algebra". *In:* G. Dutton (ed.) *Proceedings of the 1983 Harvard Computer Graphics Conference*, Vol. 2, Harvard Laboratory for Graphics and Spatial Analysis, Cambridge, Mass., pp. 127-150.

Zadeh, L.A. (1974), "Fuzzy Logic and Its Application to Approximate Reasoning". *Information Processing*, Vol. 74(3): 591-594.

Zadeh, L.A. (2002). "Some Reflections on Information Granulation and Its Centrality in Granular Computing, Computing with Words, the Computational Theory of Perceptions and Precisiated Natural Language". *In:* T. Young Lin, Y.Y. Yao and L.A. Zadeh (eds.) *Data Mining, Rough Sets and Granular Computing*, Physica-Verlag GmbH, Heidelberg, Germany, pp. 3-22.

# 3

# Towards a Quantitative Evaluation of Geospatial Metadata Quality in the Context of Semantic Interoperability

*Tarek Sboui, Mehrdad Salehi & Yvan Bédard*

## Abstract

*Semantic interoperability is a process to facilitate the reuse of geospatial data in a distributed and heterogeneous environment. In this process, the provided geospatial metadata that are appropriate for the intended use may be incomplete or not appropriate for data reuse. Thus, the external quality (fitness for use) of these metadata seems important for data reuse, since it has the potential to protect re-users from the risk of misinterpreting geospatial data. In this paper, we aim to provide a step forward in making re-users aware of the quality of geospatial metadata. We introduce a set of indicators for geospatial metadata quality and we propose a method to evaluate them with respect to the context of re-users. Based on this evaluation, we derive warnings that indicate the degree of risk of data misinterpretation related to metadata quality in the context of semantic interoperability.*

## 3.1 Introduction

Over the last decades, there has been an exponential increase in the amount of geospatial data available from multiple sources. Reusing this data can significantly decrease the cost of geospatial application (Fonseca *et al.*, 2000). In order to develop ways to enhance the reuse of available geospatial data, significant research efforts have been carried out. Among these efforts, semantic interoperability has been extensively investigated (e.g., Bishr, 1998; Harvey *et al.*, 1999), but it still remains a challenge in spite of all these efforts (Staub *et al.*, 2008). For facilitating geospatial data use/reuse, providers and application designers tag data with additional information called *geospatial metadata*. Metadata can be thematic (e.g., data acquisition method), spatial (e.g., spatial reference system used) or temporal (e.g., the time of data acquisition). The purpose of geospatial metadata is to facilitate the interpretation of data. For example, a map including a set of geometric representations (data) can be accompanied by a legend (geospatial metadata) that facilitates the interpretation of the map. However, geospatial metadata which are appropriate for a specific application may be less appropriate for another: for example, the dates of certain photographs displayed on Google Earth are often one year old, which may have no impact on several usages but may mislead others. In fact, the quality of geospatial metadata may be insufficient for certain data re-usage. Poor geospatial metadata quality may cause a risk of misinterpreting data, and may undermine the reuse of geospatial data (Agumya and Hunter, 2002), i.e., the main aim of semantic interoperability.

In order to effectively respond to the risk of misinterpreting geospatial data, re-users need to be aware of geospatial metadata quality with regard to their application. The aim of this paper is to provide a step forward in making re-users aware of the quality of metadata. We propose a method to evaluate a set of indicators for geospatial metadata quality. Such evaluation will help users to appropriately reuse geospatial data considering semantic interoperability. For example, if the quality of geospatial metadata is very low, the risk of data misinterpretation will be higher and, consequently, it may not be advisable to reuse the geospatial data.

The contributions of this paper are as follows: 1) we discuss the risk of misinterpreting data due to the lack of information about the quality of geospatial metadata in the context of semantic interoperability; 2) we introduce a set of indicators that have a major role in indicating the quality of geospatial metadata; 3) we propose a method to evaluate the quality of metadata that includes an evaluation function for each indicator and manners to calculate these functions; and 4) we use the result of quality evaluation with a framework and a set of warnings to inform re-users about the risk of data misinterpretation related to metadata quality.

In Section 3.2 we discuss the risk of data misinterpretation related to the semantic interoperability of geospatial data. Following this, we propose an evaluation of the quality of metadata to respond to the risk of data misinterpretation using appropriate warnings. We conclude and present further works in Section 3.4.

## 3.2   Semantic Interoperability of Geospatial Data and the Risk of Misinterpretation

### 3.2.1   Semantic interoperability of geospatial data

Semantic interoperability has been defined as the ability of different systems to exchange data and applications in an accurate and effective manner (Bishr, 1998; Harvey *et al.*, 1999; Brodeur, 2004). The principal aim of semantic interoperability is to reuse data and applications located in different sources.

Semantic interoperability has been viewed as the technical analogy to the human communication process (Brodeur, 2004; Kuhn, 2005). In this process, agents (a source/provider and a receiver/re-user) communicate messages (i.e., data) which are a set of organized terms. The communication process works properly when the re-user interprets data with the meaning that was originally intended by the provider (Shannon, 1948; Schramm, 1971; Bédard, 1986). To do so, re-users typically need metadata that describes the content of data regarding a specific application. For reason of simplicity, we consider that metadata consist of a set of elements. An element can be a word/phrase, a set of graphic symbols, or a combination of both. Annotation mechanisms for metadata are another key aspect but are beyond the scope of this paper. Figure 3.1 shows a snapshot of a communication process that involves only two interveners (e.g., source data producers and user's destination). Perfect communication takes place when A1 and A2 correspond exactly to each other. We should notice that such a process could also involve a chain of multiple interveners (e.g., mediators, translators, aggregators), each with their own interpretation.

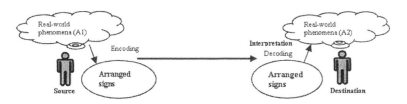

**Figure 3.1.** Sign interpretation within the communication process.

### 3.2.2   Risk of geospatial data misinterpretation related to semantic interoperability

In semantic interoperability, data may be reused in an application that is different from the one originally intended. In this process, metadata that were considered appropriate for the original application with regard to completeness, trust, clarity and levels of detail may be considered less appropriate for the second one. The appropriateness of metadata for a given application can be referred to as the external quality of metadata (fitness-for-use). Although such quality is important, evidence shows that users typically do not have a proper understanding of it (Agumya and Hunter, 2002). Consequently, they likely make wrong assumptions about the external quality of metadata. Such assumptions have the potential to expose re-users to the risk of misinterpretation. This risk is characterised by the probability of making faulty interpretations, and by the damage that can be caused by such misinterpretation.

We illustrate the risk of data misinterpretation with the following example: an agent provider sends a message, i.e., data tagged with metadata, to an agent re-user. Data consist of a set of line segments free of any intrinsic signification. We suppose the data provider is aware of eventual road-network analysis applications where data can be used and consequently tagged metadata consisting of these three elements: (1) lines represent roads, (2) bold lines indicate segments with restrictions for vehicles, and (3) road intersections have no turn restrictions unless indicated. This metadata is originally intended for a civil road network application, and is reused, by means of semantic interoperability, for an emergency road network application. Without more complete, clear or detailed indications, the semantic quality of geospatial metadata may be insufficient and lead re-users to false assumptions. This includes, for this example, that bold lines represent restrictions for all vehicles, including emergency vehicles, all line intersections are road intersections, while there could be viaducts and consequently no possibility to turn in spite of the absence of explicit restrictions in the dataset, all roads are included in the dataset while only public roads are, and so on. Such assumptions may cause a risk of faulty interpretations and usages of data. In order to effectively respond to these risks, re-users need information about the quality of metadata regarding their application.

## 3.3   Evaluating the quality of geospatial metadata in semantic interoperability

Significant research efforts have been carried out to evaluate and enhance the quality of geospatial data (e.g., Agumya and Hunter 2002; Frank *et al.*, 2004; Devillers *et al.*, 2007; Frank, 2007; Boin, 2008). In addition to geospatial data quality, reusing data in semantic interoperability requires an evaluation of metadata quality. However, data producers typically provide no quality evaluation of geospatial metadata for the intended application, nor for potential applications. In order to facilitate data reuse, an evaluation of the quality of geospatial metadata with regards to the potential applications should be proposed. Some researchers in the non-geospatial domain have studied the quality of metadata (Bruce and Hillmann, 2004; Ochoa and Duval, 2006). Although these approaches do not take into account the application in which data may be re-used, one may apply the same method to a different usage. In addition, they provide useful concepts for geospatial metadata. Nevertheless, there is still no specific work on the external quality of geospatial metadata.

In this section, we propose a restricted set of indicators and a quantitative approach for evaluating the external quality of geospatial metadata with regard to the context of re-users in a semantic interoperability process. These indicators are grouped into two categories: *global indicators* and *metadata-specific indicators*. In the first category there are three indicators – *convenience of language*, *completeness*, and *trust* – that influence the overall metadata. The second category includes one indicator that affects the quality of each metadata element, i.e. *freshness*. For each indicator, the quality value is within the interval [0, 1]. The value 1 indicates perfect quality while the value 0 indicates completely poor quality.

### 3.3.1 Quantitative evaluation of metadata

#### 3.3.1.1  Global indicators category

**Convenience of language**. This indicates the usability of a given language by those who must express and use geospatial metadata. We consider that a convenient language allows users to focus on metadata expressions, not on learning a new language. This indicator is measured according to the expressivity and the ease of understanding the language. In order to evaluate the convenience of a language, we propose a matrix (see Table 3.1) that can be used by data producers taking jointly into account both the expressivity of the language and the level of a user's knowledge of a language. For expressivity, producers can use an evaluation like the one proposed by Salehi *et al.* (2007). Then, the level of a user's knowledge of a language can be evaluated by consulting potential re-users. Knowing the expressivity and levels of user knowledge, producers can evaluate the convenience of the language using, for example, the matrix presented in Table 3.1. The values of this matrix are calculated by considering 'high' as 1, 'low' as 0 and 'medium' as 0.5. The average of intersecting columns and rows are calculated for each cell (e.g., high expressivity and medium knowledge result in $(1+0.5)/2 = 0.7$).

**Table 3.1.** Convenience of languages evaluation with respect to a user's knowledge level and language's expressivity.

|              |        | Level of user's knowledge of the language | | |
| --- | --- | --- | --- | --- |
|              |        | High | Medium | Low |
|              | High   | 1    | 0.7    | 0.5 |
| Expressivity | Medium | 0.7  | 0.5    | 0.2 |
|              | Low    | 0.5  | 0.2    | 0   |

In the example in Section 3.2.2, metadata is represented by a free natural language that has a medium expressivity when avoiding technical jargon. We suppose that the level of receiver's knowledge is high with such language. According to Table 3.2, the convenience of this language is 0.7.

**Completeness**. This indicator shows the quantity of available metadata elements with regards to the required elements. We recognize thematic, spatial, and temporal completeness. The evaluation of the completeness $P$ is calculated as:

$$P = \begin{cases} w \times \dfrac{N_{Elt}}{N_{ReqElm}} & ; if \ N_{Elt} < N_{ReqElm} \\ 1 & ; Otherwise \end{cases}$$

where $N_{Elt}$ is the number of thematic, spatial, or temporal metadata elements available, $N_{ReqElm}$ is the number of metadata elements required in a specific context, and $w$ is a predefined weight for thematic, spatial, or temporal metadata elements. This weight indicates the importance of completeness of each type in the context of reuse. If the number of available metadata elements is equal to or greater than the required elements, metadata is complete and its value is set to 1. If not, the ratio of existing elements on required elements shows the degree of completeness.

The numbers of elements $N_{ReqElm}$ and the weights can be predefined by data producers by asking potential re-users about their required metadata elements. In the example presented in Section 3.2.2 (the road-network data), the emergency application's re-users need, besides the context elements available, another element specifying the type of the obstacle. Thus, if we suppose that the weight of spatial completeness is 1, then $P = 3/4$.

**Trust**. This indicator describes the degree of faith that we have in the provided metadata. A decrease of trust lessens the external quality of metadata. Generally, in a chain of interveners, such as in semantic interoperability, if information is transmitted in a sequential manner, the trust decreases with the number of interveners (Bédard, 1986; Moe and Smite, 2007). We evaluate the trust using the following function:

$$T = \frac{\sum \alpha_i}{N}$$

where $\alpha_i$ is the confidence given to the $i^{th}$ intervener. The value of confidence is between 0 and 1. $N$ is the number of interveners that transmitted the metadata element. We consider that each intervener transmits a metadata element just once. In the provided example in Section 3.2.2, we suppose that the metadata of roads network went through three interveners in a semantic interoperability chain. The first has a confidence of 0.8, the second has a confidence of 0.5 and the third has a confidence of 0.8. Consequently, $T = (0.8+0.5+0.8)/3 = 0.7$.

At this point, one must remember that these indicators do not aim at being complete or precise, but solely at indicating if potential problems may arise without having to make a complete analysis containing all details. Such a method is frequently used in complex domains involving large numbers of hard-to-measure factors, such as the fields of epidemiology, ecology, economics, etc. It has already been used in the evaluation of geospatial data quality by Devillers et al. (2007) and the proposed selection of indicators can be modified according to the context at hand if needed.

### 3.3.1.2 Metadata-specific indicators category

**Freshness**. This quality indicator shows the degree of freshness related to the use of a metadata element at a given time with regard to its lifetime. Accordingly, the value of freshness is determined by the age and lifetime of the metadata. In semantic interoperability, the age of the metadata element $e$ is the difference between the time of interpretation $T_{int}(e)$ and the time of the definition of that context element $T_{def}(e)$. The freshness of the context element $F(e)$ is evaluated as follows:

$$F(e) = \begin{cases} 1 - \dfrac{T_{int}(e) - T_{def}(e)}{lifetime(e)} ; & if \quad T_{int}(e) - T_{def}(e) < lifetime(e) \\ 0 \quad ; & Otherwise \end{cases}$$

where *lifetime* is the expected period of time after which a metadata element is no longer meaningful. The freshness value of context element $F(e)$ decreases as its age increases. A low value of freshness has a negative impact on the quality of metadata element. Lifetime and the time of definition $T_{def}$ can be introduced by a data producer. In the road-network data example (Section 3.2.2), we suppose that the metadata element "bold lines indicate restrictions for vehicles" was defined in the year 2000 when the dataset was created, and will be obsolete after 30 years. Then $F(e) = 1 - (2008-2000/30) = 0.73$.

### 3.3.2 A framework for evaluating the external quality of metadata

In this section, we present a framework to evaluate the external quality of metadata. As shown in Figure 3.2, this framework consists of two layers: the detailed layer and the global layer. The detailed layer is devoted to evaluate the quality of metadata elements based on the freshness indicator. The global layer consists of two parts. The first part is derived from the detailed layer by calculating the weighted average of the qualities of metadata elements. The second part presents the quality evaluation of overall metadata based on the convenience, completeness and trust indicators. Such global and detailed representation allows re-users to get a global picture (quality of overall geospatial metadata), and then dig into the quality evaluation of each metadata element when needed to have a detailed perspective. The approach of considering different layers for quality analysis bears similarities with the method proposed by Devillers *et al.* (2007).

In each layer, while good metadata quality indicates it is less likely to have a risk of data misinterpretation, a poor quality indicates a higher risk. In order to help re-users to intuitively understand the degree of such a risk, we provide a set of warnings which are based on standard danger symbols proposed by ANSI Z535.4 (1991) and previously used in the geospatial domain by Levesque *et al.* (2007). These symbols show the degree of the risk of data misinterpretation related to metadata quality at each layer, and thereby stimulate appropriate responses to such a risk. For example, if the warning is 'Danger', it would be better not to reuse geospatial data by means of semantic interoperability. Table 3.2 shows an example of how warnings can be predefined with regards to quantitative values of geospatial metadata quality.

**Table 3.2.** An example of indicators based on the geospatial metadata quality.

| Quality metadata | Indicator of risk of data misinterpretation |
|---|---|
| Q <0.2 | ⚠ DANGER |
| 0.2 < = Q <0.5 | ⚠ WARNING |
| 0.5 < = Q <0.75 | ⚠ CAUTION |
| Q >= 0.75 | NOTICE |

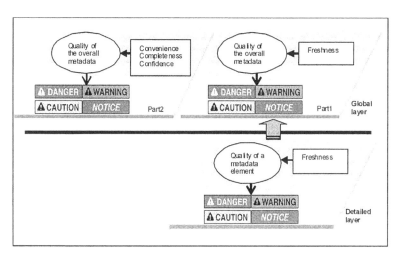

**Figure 3.2.** A framework for evaluating the external quality of metadata.

Referring to our example, in the detailed layer, the freshness of metadata element "bold lines indicate restrictions for vehicles" is 0.73; thus, for this element the 'caution' indicator has to be shown to the re-user. In order to calculate the freshness in the part one of the global level, we suppose that for two other elements of the metadata the freshness is 0.5 and 0.1. Thus, if the weight of these three elements is 1, the freshness of the overall metadata is the average of three values 0.73, 0.5, and 0.1, i.e., 0.44. The 'warning' indicator has to be shown to the re-user to indicate the risk associated to this part of global level. In the second part of the global level, the completeness is 0.75; thus, the 'notice' indicator has to be displayed.

## 3.4  Conclusion

Reusing geospatial data by means of semantic interoperability faces a risk of misinterpretation. This risk may be due to the fact that geospatial metadata, which is appropriate for the intended use, may be of poor external quality for data reuse. Evaluating this quality can help protect re-users from the risk of misinterpreting geospatial data. In this paper, we proposed a method to evaluate the external quality of geospatial metadata with respect to data re-users. For that, we introduced a set of indicators and quantitatively evaluated the quality of geospatial metadata. These indicators are organised into a framework consisting of two layers: the detailed layer and global layer. This framework allows re-users to get a global picture, and then drill down for the evaluation of the quality of each geospatial metadata element. In addition, we proposed a set of warnings that inform re-users about the risk of data misinterpretation related to metadata quality.

We should note that these indicators do not aim at completely eliminating the risk of data misinterpretation and reuse in the context of semantic interoperability, but rather represent a step forward in making re-users aware of such a risk. Further research is underway to define additional indicators such as relevancy and granularity of metadata elements. Then, the proposed approach will be implemented to show how the proposed indicators enhance semantic interoperability of geospatial data.

## Acknowledgments

The authors wish to thank for its support the NSERC Industrial Research Chair of Geospatial Database for Decision Support financed by the Natural Sciences and Engineering Research Council of Canada, Laval University, Hydro-Québec, Research and Development Defence Canada, Natural Resources Canada, Ministère des Transports du Québec, KHEOPS Technologies, Intélec Géomatique, Syntell, Holonics, and DVP-GS. Also, we would like to thank the reviewers for their valuable comments.

## References

Agumya, A. and. G.J. Hunter. (2002), "Responding to the consequences of uncertainty in geographical data". *International Journal of Geographical Information Science*, Vol 16(5): 405-417.

ANSI (1991), *American National Standard for Product safety signs and labels*, ANSI Z535.4.

Bédard, Y. (1986), *A Study of the Nature of Data Using a Communication-based Conceptual Framework of Land Information*. PhD thesis, University of Maine, USA, 260p.

Bishr, Y. (1998), "Overcoming the semantic and other barriers to GIS interoperability", *International Journal of Geographical Information Science*, Vol. 12(4): 299-314.

Boin, A. (2008), *Exposing Uncertainty: Communicating spatial data quality via the Internet*. PhD thesis, Department of Geomatics, The University of Melbourne, Australia, 183p.

Brodeur, J. (2004), *Interopérabilité des données géospatiales: élaboration du concept de proximité géosémantique*. PhD thesis, University of Laval, Canada, 267p.

Bruce, T.R. and D.I. Hillmann. (2004), "The continuum of metadata quality: defining, expressing, exploiting". *In:* D.I. Hillman and E.L. Westbrooks (eds.) *Metadata in Practice*, American Library Association, Chicago, US, pp. 238-256.

Devillers, R., Y. Bédard, R. Jeansoulin. and B. Moulin. (2007), "Towards spatial data quality information analysis tools for experts assessing the fitness for use of spatial data". *International Journal of Geographical Information Science*, Vol. 21(3): 261-282.

Fonseca, F., M. Egenhofer, C. Davis, and K. Borges. (2000), "Ontologies and Knowledge Sharing in Urban GIS". *Computers, Environment, and Urban Systems*, Vol 24(3): 232-251.

Frank, A.U. (2007), "Assessing the quality of data with a decision model". *In:* M. Molenaar, W. Kainz and A. Stein (eds.) *Modelling Qualities in Space and Time. Proceedings of the 5th International Symposium on Spatial Data Quality '07, ITC, Enschede, The Netherlands*. http://www.itc.nl/ISSDQ2007/proceedings/index.html.

Frank, A.U., E. Grum and B. Vasseur. (2004), "Procedure to select the best dataset for a task". *In:* M.J. Egenhofer, C. Freksa and H.J. Miller (eds.) *GIScience 2004*. LNCS, Vol. 3234: 81-93.

Harvey, F., W. Kuhn, H. Pundt, Y. Bishr and C. Riedemann. (1999), "Semantic Interoperability: A Central Issue for Sharing Geographic Information". *The Annals of Regional Science (Special Issue on Geo-geospatial Data Sharing and Standardization)*, Vol. 33(2): 213-232.

Kuhn, W. (2005), "Geospatial Semantics: Why, of What, and How". *Journal on Data Semantics (Special Issue on Semantic-based GIS)*, LNCS, Vol. 3534: 1-24.

Levesque, M.A., Y. Bédard, M. Gervais and R. Devillers. (2007), "Towards managing the risks of data misuse for spatial datacubes". *In:* M. Molenaar, W. Kainz and A. Stein (eds.) *Modelling Qualities in Space and Time. Proceedings of the 5<sup>th</sup> International Symposium on Spatial Data Quality '07, ITC, Enschede, The Netherlands.* http://www.itc.nl/ISSDQ2007/proceedings/index.html.

Moe, N.B. and D. Smite. (2007), "Understanding Lacking Trust in Global Software Teams: A Multi-Case Study". *In:* J. Münch and P. Abrahamsson (eds.) *Product-Focused Software Process Improvement*, LNCS, Vol. 4589: 20-32.

Ochoa, X. and E. Duval. (2006), "Towards automatic evaluation of learning object metadata quality". *In:* J.F. Roddick, R. Benjamins, S. Si-Said Cherfi, R. Chiang, R. Elmasri, H. Han, M. Hepp, M. Lystras, V. Misic, G. Poels, I.-Y. Song, J. Trujillo and C. Vangenot (eds.) *Advances in Conceptual Modeling Theory and Practice*, LNCS, Vol. 4231: 372-381.

Salehi M., Y. Bédard, M.A. Mostafavi and J. Brodeur. (2007), "On Languages for the Specification of Integrity Constraints in Spatial Conceptual Models". *In:* J.-L. Hainaut, E.A. Rundensteiner, M. Kirchberg, M. Bertolotto, M. Brochhausen, P. Chen, S. Sisaid Cherfi, M. Doerr, H. Han, S. Hartmann, J. Parsons, G. Poels, C. Rolland, J. Trujillo, E. Yu and E. Zimlanyi (eds.) *Advances in Conceptual Modeling – Foundations and Applications*, LNCS, Vol. 4802: 388-397.

Schramm, W. (1971), "How Communication Works". *In*: J.A DeVito (ed.) *Communication: Concepts and Processes*. Prentice-Hall Inc., Englewood Cliffs, New Jersey, pp. 12-21.

Shannon, C.E. (1948), "A Mathematical Theory of Communication". *The Bell System Technical Journal*, Vol. 27: 379-423.

Staub, P., H.R. Gnagi and A. Morf. (2008), "Semantic Interoperability through the Definition of Conceptual Model Transformations". *Transactions in GIS*, Vol. 12(2): 193-207.

# 4

## Characterizing and Comparing GIS Polygon Shape Complexity by Iterative Shrinking Spectra

*Tarmo K. Remmel, Connie Ko & Rick L. Bello*

## Abstract

*Shape characterization and comparison is often necessary wherever vector polygons are used to partition space among regions with different attributes. From an ecological perspective, the shape of land cover patches may influence habitat selection or ecosystem processes while in anthropogenic settings, property shapes may be differentially valued. The current literature provides several means for characterizing polygon shape; unfortunately, these are typically single number summaries which disregard the original complexity of the shapes they intend to measure. The method demonstrated in this paper imposes an incremental shrinking (i.e., internal buffering) to vector polygons interleaved with the computation of summary parameters to provide a continuum of shape characterizations along an axis of finite length having as many data points as there are shrinking operations. The connection of these data points constructs a one dimensional spectrum that details the characteristics (e.g., area, perimeter, parts) of the shape it represents. The ShrinkShape technique provides a multi-dimensional, scalable, rotation-invariant, and detailed approach for the comparison of vector polygon shapes that currently cannot be achieved by approaches relying on single number summaries. We illustrate the technique with synthetic (fabricated) shapes to highlight widely varying phenomena. Further classifications of these spectra provide means for grouping spatial features with similar shape complexities.*

## 4.1 Introduction

A natural system when observed at time $t_0$ will possess emergent characteristics that include scale dependent spatial patterns. These spatial patterns will reflect the state of that system at the time of observation but will also be partly influenced by decisions made regarding the system's representation (Moody and Woodcock, 1995). When the same system is observed through time ($t_0 \rightarrow t_n$), these patterns become dynamic and change to reflect the result of complex interacting processes that affect that system. Some landscape ecologists have sought to link these processes with observed patterns within an inferential framework (cf. Fortin *et al.*, 2003) in hopes that, in specific cases, observations of patterns (realizations of spatial processes) could be used to make inferences regarding the underlying processes that yielded those patterns. This goal of linking pattern and process has extensive support, especially in the ecological and landscape ecological literature (Sarnelle, 1994; Carlton, 1996; Nagendra *et al.*, 2003; Stallins and Parker, 2003; Gardner and Urban, 2007).

The characterization, quantification, and comparison of spatial patterns interests geographers and ecologists as demonstrated by fruitful publication within this domain (Fortin *et al.*, 2003; Turner, 2005; Boots, 2006). Numerous (and often redundant) spatial metrics (often referred to as landscape pattern indices – LPI) have been devised and promoted (Riitters *et al.*, 1995; Haines-Young and Chopping, 1996; Gustafson, 1998); their proliferation has resulted primarily due to readily accessible software to facilitate their computation (Baker and Cai, 1992; McGarigal and Marks, 1995), but not without difficulties, uncertainty, and error (Remmel and Csillag, 2003; Li and Wu, 2004).

While most LPI characterize individual elements of spatial pattern (e.g., size, shape, connectivity, and contrasts), an holistic and robust measure of spatial pattern has yet to be presented, tested, and proved useful. Most LPI provide results in non-transferable units and thus composition and configuration are typically measured in incompatible units. Remmel and Csillag (2006) proposed an innovative technique for comparing compositional and configuration similarities/differences between categorical raster maps based on mutual information spectra (using common measurement units – bits). While this method is effective for comparing categorical landscapes, it provides little information regarding the shape of image entities and cannot be ported into a vector GIS environment.

Shape of individual entities however remains a key characteristic in theoretical object recognition (Henderson and Davis, 1981; Belongie *et al.*, 2002; Zhang *et al.*, 2003; Shokoufandeh *et al.*, 2006), and is also largely used while interpreting aerial photography, satellite imagery, or performing other geographic spatial analyses (Lee and Sallee, 1970; Arnold, 1996; Antrop and Van Eetvelde, 2000). While several indices to quantify shape have been proposed, the perimeter-to-area ratio (PA) (Li and Narayanan, 2003) is likely the most common, but also highly influenced by the size of the object that is being measured. For this reason, the use of the corrected perimeter-to-area ratio (cPA) is preferred (Antrop and Van Eetvelde, 2000), as it provides a quantity relative to either a reference square (useful for raster data) or circle (useful for vector data) having the same area as the shape in question. Lee and Sallee (1970) prove a theorem that states *"there exists no continuous one-to-one function from S, the set of all plane shapes, into R, the set of real numbers"* and rather compare entities to known standard shapes (e.g., circle, square, or rectangle). Other approaches have sought to classify shapes based on their compactness (cf. MacEachren, 1985), sometimes by counting contact perimeters for entities comprised of cells and comparing the number of internal cell edges of a shape to a theoretical maximally compact shape comprising an identical number of cells (Bribiesca and Guzmán, 1979; Bribiesca, 1997).

A difficulty with many shape metrics is that they typically summarize complex entities with a single value, although some exploration into multiple-value summaries do exist (Mehtre *et al.*, 1997; Schuldt *et al.*, 2006). Zhang and Lu (2004) summarize several common shape representation and description techniques, and identify the importance of metrics being translation and scale invariant.

Our proposed measurement and characterization technique for planar shapes, *ShrinkShape*, provides a repeatable method that is translation, rotation, and (potentially) scale invariant, while providing informative information regarding several dimensions of shape description. We define the devised algorithm and demonstrate its use with synthetic shapes.

## 4.2  Methods

Our algorithm, ShrinkShape, runs with the ArcGIS ModelBuilder to be compatible with Shapefiles (ESRI, 1998). The algorithm is intended to work with any data that can be represented by vector polygons distributed on a two-dimensional plane – the algorithm requires no information regarding context or attributes. Typically, this means capturing the perimeters (outlines) of geographic phenomena and representing them by enclosing areas with lines of varying complexity, as determined by the number and distribution of vertices. Each enclosed area is considered as a unique shape indexed by an integer ($i$), and is processed independently but simultaneously with all shapes within a single Shapefile. The only exception to this base rule is the inclusion of any island polygons within the context of a larger shape; these cases are all treated simultaneously as a single perforated shape. For example, a lake polygon that contains an island will be considered as a lake polygon having a hole; the land component does not form part of the water component.

Since the algorithm repeatedly buffers each shape internally (thereby shrinking the polygon) using a consistent and specified distance ($d$), this 1-dimensional parameterization is the only information required to run the algorithm, other than supplying the spatial data containing the polygons of interest. Prior to each shrinking phase (indexed by s), three summary measurements are made for each unique shape; these include total area ($A_{is}$), total perimeter ($P_{is}$), and a total number of child parts ($N_{is}$) existing from the original parent shape. While the concept of area and perimeter are self-explanatory, the concept of the number of parts is elaborated further below.

Prior to any shrinking, each unique polygon (shape) can be considered as a parent. After an internal buffering operation (shrinking phase), the result for a given parent polygon will be represented by one or more child shapes, depending on whether the shrinking operation pinches off any original lobes. Whether part of the polygon will be pinched off is a direct function related to edge complexity, existence of narrow regions within the shape, and the degree of shrinking. Thus, if a parent shape which looks like an hourglass is shrunk, at some point the shape will split into two parts, disconnected in the middle, to form two child parts for the original single parent. The area and perimeter of these child parts are summed and included in the total area and total perimeter accounting at each shrinking phase.

The alternating task of measuring $A_{is}$, $P_{is}$, and $N_{is}$, with the shrinking of each unique shape by $d$ continues iteratively until each shape ceases to exist. This happens when each child part is forced to shrink beyond the spatial limits that would allow it to retain some area and a perimeter. Once a shape has been eliminated, it no longer contributes to new measures of $A_i$, $P_i$, or $N_i$ (they are all zero), but any remaining entities in the Shapefile will continue to be shrunk until all shapes are eliminated. Ultimately, the spectra of $A_i$, $P_i$, and $N_i$ are plotted along the continuum of shrinking phases to produce plots depicting the change in the measured parameters relative to the degree of shrinking. Such spectra for $A_i$ and $P_i$ decrease monotonically, while this is not necessarily true for $N_i$.

We demonstrate the characteristics of this algorithm using synthetic shapes (Figure 4.1) that illustrate key behaviours. These include demonstrations with classical simple geometric shapes (Figure 4.1a-d) and more complex shapes (Figure

4.1d-h) to illustrate rotation invariance, scale robustness, and the ability to compare the compactness of shapes.

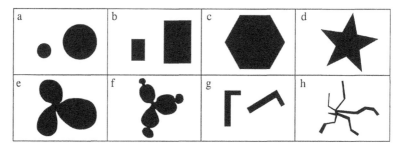

**Figure 4.1.** Representative synthetic shapes used to demonstrate the algorithm. All shapes are relatively sized and oriented. Shape descriptors include: a) circles, b) rectangles, c) hexagon, d) star, e) lobes, f) beads, g) L-shaped, and h) sinuous.

## 4.3 Results

Our analysis was conducted on numerous shapes including circles, rectangles, hexagons, stars, lobes, connected beads, L-shapes, and several synthetic sinuous forms as illustrated in Figure 4.1. The first result is the most obvious; large, but compact shapes require more shrinking phases than do shapes with lower maximum thicknesses while $d$ is held constant. Thus a square and a highly sinuous shape with identical areas will not require an identical number of shrinking steps to fully characterize their ShrinkShape spectra. The second most obvious result is that two shapes (e.g., two circles), differing only by their size (i.e., one is scaled larger than the other with an identical number of vertices) will require a different number of shrinking steps to complete the operation. Thus, the number of steps required to completely shrink the shape to the point of elimination is governed by the maximum thickness ($T_m$) of the shape, not its maximum width ($W_m$), where $T_m$ can be described as the radius of the largest circle than can be drawn within the shape. The maximum number of shrinking iterations (n) for a given shape is found by dividing the maximum thickness by $2 \cdot d$ (twice the shrinking distance).

For perfectly convex shapes, the resultant shrinking spectrum for area will decrease according to one limb of a second-order polynomial function ($R^2 = 1.0$); while shapes deviating from the perfect convex case will also begin to deviate from the second-order polynomial trajectory that describes the trend of decreasing area with shrinking phase. Perimeters for perfect convex shapes will decrease linearly ($R^2 = 1.0$) while non-convex shapes will express irregular, but monotonically decreasing functions. The functional form for the number of child parts will vary based on the number and distribution of pinch-points (regions where convexity occurs to significantly reduce the local thickness of a shape relative to adjacent lobes. The form of the spectrum is not monotonic and will always begin at 1. Depending on the degree of concavity (e.g., stars); shrinking may never produce $N_{is} > 1$.

While certain families of synthetic shapes behave similarly, only representative results will be provided to avoid redundancy and to conserve space. Figure 4.2

illustrates the shrinking spectra results for the hexagon shape (as defined in Figure 4.1c) for shrinking steps s = 1 ... s = (n - 1). Since this is a perfectly convex shape, the perimeter spectra form a perfectly decreasing straight line (y = -69.282x + 2050.2, $R^2$ = 1.0000), while the area spectra decrease monotonically along a $2^{nd}$ order polynomial function (y = 0.0346x$^2$ - 2.0502x + 30.336, $R^2$ = 1.0000) until shrinking phase n - 1. The final phase is not plotted on these graphs or included in these equations due to the remainder effect introduced in the final shrinking phase prior to shape extinction. Since equal multiples of *d* are unlikely to divide perfectly into the functional shape thickness, the final phase must be considered separately if these spectra are to be described by smooth functional forms. The results for the Hexagon reflect those of several other synthetic convex shapes (i.e., circles, triangles, rectangles, and octagons), while the equations will differ to reflect the sizes of the relative shapes. Given that convex shapes have no pinch-points (no part of the shape may express concavity), regardless of the magnitude selected for *d*, the original shape is never severed into smaller child parts and thus the spectra for parts remain constant at 1 (and obviously will drop to zero at phase n, when the shape becomes extinct).

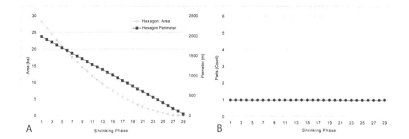

**Figure 4.2.** Hexagon shrinking spectra for area (A), perimeter (A), and the number of child parts (B) for shrinking phases 1...(n - 1).

In the examination of non-convex shapes (e.g., lobes, depicted in Figure 4.1e), we introduce pinch-points, where the boundary of the shape inflects inward, causing a local reduction in the shape thickness and providing locations at which shrinking can orphan areas of the parent shape. Shrinking spectra for the lobes shape are provided in Figure 4.3. In Figure 4.3a, the area spectra look very similar to that of convex shapes, but when a function is fitted to these points, the relationship is not a perfect $2^{nd}$ order polynomial, albeit very close (y = 0.0259x$^2$ - 0.9159x + 8.1891, $R^2$ = 0.9996). More interestingly, the perimeter spectrum deviates visibly from a linear form and begins to illustrate key components of the lobes shape. Notice that, between shrinking phases 5 and 6, the perimeter drops drastically relative to the first 5 phases. While the perimeter spectrum is linear, the shape is shrinking as a convex shape, once a pinch-point is reached, the relationship between area, parts, and perimeter alter drastically and, depending on the configuration of the shape's lobes, the length of perimeter can quickly decrease. Once the major lobes are split into multiple parts, they too begin to shrink as a series of convex shapes, yielding a negative linear trajectory to the perimeter spectrum. This process continues until the shape is shrunk to extinction.

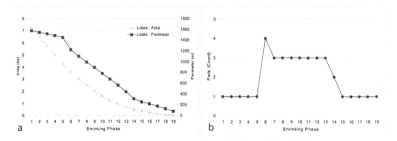

**Figure 4.3.** Lobes shrinking spectra for area (a), perimeter (a), and the number of child parts (b) for shrinking phases 1...(n - 1).

Figure 4.3b provides the spectrum for the number of parts during the shrinking phases of lobes. Of note is the drastic increase from 1 to 4 parts while moving from phase 5 to 6. This indicates that at this point, the original lobes shape is split into 4 distinct parts (i.e., the 3 main lobes and a small central polygon). The small central polygon is quickly eliminated during the subsequent shrinking phase and thus the parts count drops to 3; at phases 14 and 15 two more lobes are expired and the last lobe disappears at phase 20. Together, the perimeter and parts count spectra illustrate the general form of how area is distributed within the shape, giving insight to pinch-points and thus edge complexity.

As seen in the previous example with the lobes, shapes that diverge from convexity begin to express area, perimeter, and parts spectra that deviate from $2^{nd}$ order polynomial, linear, and uniform theoretical distributions. A second example of this would be from Figure 4.2h (Sinuous), whose results are provided in Figure 4.4. The area spectrum ($y = 0.0504x^2 - 2.1214x + 22.255$, $R^2 = 0.9997$) deviates from the theoretical and the perimeter spectrum indicates numerous pinch-points and alterations in the rate of perimeter decrease. The edge complexity is high and there are few extended ranges where the perimeter spectrum is linear, thus, there appear to be no large and dominant convex areas comprising this shape.

The parts spectrum for our synthetic sinuous shape (Figure 4.4b) indicates the number of pinch-points, child-part extinctions, and relative shape complexity. Since the only prolonged phase of uniformity with regard to the number of parts is at the beginning of the shrinking phases, it can be deduced that although there is complexity within the shape, these complexities are secondary and that the shape exhibits some minimum thickness throughout. Each increasing segment on the parts spectrum identifies the presence of a pinch-point (or simultaneous pinch-points), while falling limbs indicate part extinctions. The duration (number of phases) that the function remains at some level indicates the relative size (area) of the parts and can be used to assess their dominance.

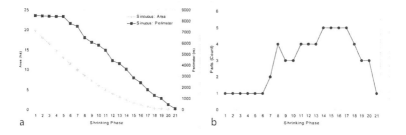

**Figure 4.4.** Sinuous shrinking spectra for area (a), perimeter (a), and the number of child parts (b) for shrinking phases 1...(n - 1).

A favourable property of the ShrinkShape technique is that it is rotation invariant; that is, regardless of the orientation of a shape, the produced spectra for area, perimeter, or parts will always be identical. This approach is not coordinate system specific and thus any rotation of a shape on a Cartesian plane will possesses this property. To demonstrate, we use the L-shaped objects from Figure 4.1g and produce their respective area, perimeter, and parts' spectra (Figure 4.5); note that due to its uniform nature, we do not show the spectrum for shape parts. The L-shaped objects are not perfectly convex and thus their area spectra deviate slightly from the theoretical $2^{nd}$ order polynomial form ($y = 241.52x^2 - 11613x + 104108$, $R^2 = 0.9996$), yet the two spectra overlap perfectly and exhibit identical form. Similarly, the perimeter spectra are identical and express a shape with two relatively dominant and compact regions. However, these rotated shapes lack a significant pinch-point and thus fail to produce additional child parts at any point along the shrinking continuum. Figure 4.5 illustrates the power of ShrinkShape to characterize shapes in any orientation, but also indicates the inability to capture orientation effects or information regarding spatial distribution (requirements needing additional spatial tools).

A second major consideration in measuring shape is spatial scale. While Shrink-Shape might be rotation invariant, it can be sensitive to spatial scale; two shapes differing only by enlargement or reduction will produce different output spectra. However, these spectra will simply be scaled versions of each other. To demonstrate, we took the original lobe shape and both enlarged and reduced it, resulting in three sizes of the identical shape (Figure 4.6). We plotted the perimeter spectra for each of the three scaled shapes (Small, Medium, and Large) where the differences attributed to changing values of $T_m$, since n differs among each and $d$ was held constant. Trends and form of these spectra are the same.

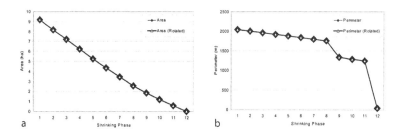

a                                                    b

**Figure 4.5.** Graphs depicting the rotation invariance among the area (a) and perimeter (b) spectra for the synthetic L-shaped items depicted in Figure 4.2g. The
spectra for the two identical (but rotated) shapes are provided for shrinking phases
1…n. Note that the spectra for both shapes are identical and that the spectrum for
child parts (not shown) is similarly rotation invariant.

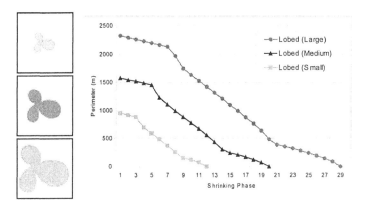

**Figure 4.6.** Perimeter spectra for small, medium, and large scaled versions of the lobed
shape to illustrate the effect of scaling but the consistency in functional form among the
three versions of the identical shape.

To compensate for scaling effects, it is possible to measure $T_m$ for individual
shapes and then, by fixing the maximum number of shrinking phases, compute a
unique distance $d$ for each shape. This will ensure that each spectrum has precisely
the same number of steps. Once the values for the spectra (area, perimeter) have
been computed, they can be normalized by dividing each value along the individual
spectra by the total area or perimeter to obtain a ratio which can be plotted and
compared among shapes. Figure 4.7 illustrates this normalized ratio technique for
three scaled perimeters of a small lake (each shape has an identical number of vertices and differ only in their scaled size). The three normalized ratio perimeter spectra overlap completely; thus, allowing the three shapes to be considered identical.
This provides a meaningful means of assessing shape equality or similarity, often
necessary for object recognition or data quality assessments.

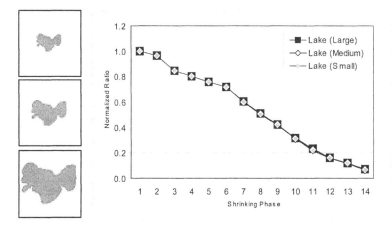

**Figure 4.7.** Superimposed spectra from three identical shapes that have been scaled to different sizes. The spectrum values have been normalized with an equal number of shrinking phases to ensure comparability.

Our investigations continued to be plagued by the question: how sensitive is the observed spectra to the selection of $d$ and how can its selection be optimized? To investigate this issue, we performed the ShrinkShape procedure on a sample of arctic pond delineations with multiple $d$ values. The results were plotted in three-dimensional space (one example is provided in Figure 4.8) where the x-axis represents the total cumulative shrinking distance, the y-axis represents perimeter, and the z-axis represents area (thus ignoring the parts component). The plots indicate that regardless of the value $d$ used (ranging from 5m to 50m in 5m increments), the spectra are coarsely identical in their form; thus being a strong indicator of the scaling resistance of the method and the flexibility in selecting $d$. When examined more closely (see the inset in Figure 4.8), fine-scale variability is noted among the spectra, indicative of subtle changes due to changes in the number of vertices along each line. This result reflects that while the selection of $d$ does have an impact on the fine form of these spectra, the absolute value is not overly critical for the general description of the planar shape. The importance of this finding is that while a sufficient number of shrinking iterations are maintained (e.g., 10-20), the absolute magnitude of $d$ will not cause significant deviation in the form of the spectra.

## 4.4   Discussion and Conclusions

We demonstrate an elegant and robust method for characterizing planar shapes based on computing spectra of summary variables using an iterative shrinking algorithm. These spectra capture both the edge complexity of the shapes and characteristics of their form, which can then be compared between or among multiple shapes. Our examples illustrate the invariance due to shape orientation (e.g., rotations) and can thus be used for object recognition or comparisons. Additionally, spectra can indicate convexity, concavity, and pinch-points, all characteristics that define the irregularity and compactness of shapes. Modification of the shrinking distance parameter further permits the algorithm to be considered resistant to scaling issues, whereby two (or more) identical shapes differing only by a spatial scaling factor can be identified as being the same.

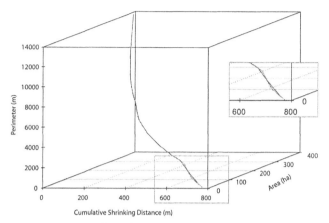

**Figure 4.8.** A three-dimensional plot of ShrinkShape results for a single Arctic pond delineation. The multiple spectra represent $d$ values varied from 5m to 50m with 5m increments. The results indicate a consistent spectral form across all application of $d$ with slight local deviations as seen in the inset diagram.

In terms of spatial data quality, planar shapes can be compared by their Shrink-Shape spectra at varying levels of spatial detail (i.e., changes to the $d$ parameter), where at coarse spatial detail (large $d$) shapes have similar spectra but differ at fine spatial detail (small $d$). The complexity of a shape's perimeter is ultimately captured in the irregularity of the ShrinkShape spectra and can be used to assess both the convexity and complexity of boundaries at multiple scales. For purposes of shape comparison, shape matching, complexity assessment, or the comparison of multiple digitization efforts for a common shape, ShrinkShape offers a repeatable and objective means to quantify, represent, and assess these conditions.

While similar spectra represent structurally similar planar shapes, deviations indicate differences. Therefore, these spectra can be used to identify degrees of uncertainty among shapes and potentially provide a useful tool for fuzzy-matching or structural similarity selection operations. The proposed technique illustrates a novel method for representing planar shape by a series of spectra that maintain more detail than typical single-number summaries and allow for evaluations otherwise impossible or significantly more difficult. Our planned extension of this theoretical work is to link these shape spectra with physical processes acting upon natural landscapes, thus, starting to draw a link between patterns and processes.

## Acknowledgments

We acknowledge the financial support of the Natural Sciences and Engineering Research Council of Canada.

## References

Antrop, M. and V. Van Eetvelde. (2000), "Holistic aspects of suburban landscapes: visual image interpretation and landscape metrics". *Landscape and Urban Planning*, Vol. 50(1-3): 43-58.

Arnold, R.H. (1996), *Interpretation of airphotos and remotely sensed imagery*, Prentice Hall, Upper Saddle River, NJ, 249p.

Baker, W.L. and Y.M. Cai. (1992), "The r.le-programs for multiscale analysis of landscape structure using the GRASS geographical information-system". *Landscape Ecology*, Vol. 7(4): 291-302.

Belongie, S., J. Malik, and J. Puzicha. (2002), "Shape matching and object recognition using shape contexts". *IEEE Transactions on Pattern Analysis and Machine Intelligence*, Vol. 24(4): 509-522.

Boots, B. (2006), "Local configuration measures for categorical spatial data: binary regular lattices". *Journal of Geographical Systems*, Vol. 8(1): 1-24.

Bribiesca, E. (1997), "Measuring 2-D shape compactness using the contact perimeter". *Computers & Mathematics With Applications*, Vol. 33(11): 1-9.

Bribiesca, E. and A. Guzmán. (1979), "Shape description and shape similarity measurement for two-dimensional regions". *Geo-Processing*, Vol. 1(2): 129-144.

Carlton, J.T. (1996), "Pattern, process, and prediction in marine invasion ecology". *Biological Conservation*, Vol. 78(1-2): 97-106.

ESRI. (1998), *ESRI Shapefile technical description: an ESRI white paper*, ESRI, Redlands CA, 28p.

Fortin, M.J., B. Boots, F. Csillag and T.K. Remmel. (2003), "On the role of spatial stochastic models in understanding landscape indices in ecology". *Oikos*, Vol. 102(1): 203-212.

Gardner, R.H. and D.L. Urban. (2007), "Neutral models for testing landscape hypotheses". *Landscape Ecology*, Vol. 22(1): 15-29.

Gustafson, E.J. (1998), "Quantifying landscape spatial pattern: what is the state of the art?". *Ecosystems*, Vol. 1(2): 143-156.

Haines-Young, R. and M. Chopping. (1996), "Quantifying landscape structure: a review of landscape indices and their application to forested landscapes". *Progress in Physical Geography*, Vol. 20(4): 418-445.

Henderson, T.C. and L. Davis. (1981), "Hierarchical models and analysis of shape". *Pattern Recognition*, Vol. 14(1-6): 197-204.

Lee, D.R. and G.T. Sallee. (1970), "Method of measuring shape". *Geographical Review*, Vol. 60(4): 555-563.

Li, J. and R.M. Narayanan. (2003), "A shape-based approach to change detection of lakes using time series remote sensing images". *IEEE Transactions on Geoscience and Remote Sensing*, Vol. 41(11): 2466-2477.

Li, H.B. and J.G. Wu. (2004), "Use and misuse of landscape indices". *Landscape Ecology*, Vol. 19(4): 389-399.

MacEachren, A.M. (1985), "Compactness of geographic shape – comparison and evaluation of measures". *Geografiska Annaler Series B-Human Geography*, Vol. 67(1): 53-67.

McGarigal K. and B.J. Marks. (1995), *FRAGSTATS: spatial pattern analysis program for quantifying landscape structure*. U.S. Department of Agriculture, Forest Service, Pacific Northwest Research Station. Gen. Tech. Rep. PNW-GTR-351, 122p.

Mehtre, B.M., M.S. Kankanhalli and W.F. Lee. (1997), "Shape measures for content based image retrieval: A comparison". *Information Processing & Management*, Vol. 33(3): 319-337.

Moody, A. and C.E. Woodcock. (1995), "The influence of scale and the spatial characteristics of landscapes on land-cover mapping using remote sensing". *Landscape Ecology*, Vol. 10(6): 363-379.

Nagendra, H., J. Southworth and C. Tucker. (2003), "Accessibility as a determinant of landscape transformation in western Honduras: linking pattern and process". *Landscape Ecology*, Vol. 18(2): 141-158.

Remmel, T.K. and F. Csillag. (2003), "When are two landscape pattern indices significantly different?". *Journal of Geographical Systems*, Vol. 5(4): 331-351.

Remmel, T.K. and F. Csillag. (2006), "Mutual information spectra for comparing categorical maps". *International Journal of Remote Sensing*, Vol. 27(7): 1425-1452.

Riitters, K.H., R.V. O'Neill, C.T. Hunsaker, J.D. Wickham, D.H. Yankee, S.P. Timmins, K.B. Jones and B.L. Jackson. (1995), "A factor analysis of landscape pattern and structure metrics". *Landscape Ecology*, Vol. 10(1): 23-39.

Sarnelle, O. (1994), "Inferring process from pattern – trophic level abundances and imbedded interactions". *Ecology*, Vol. 75(6): 1835-1841.

Schuldt A., B. Gottfried and O. Herzog. (2006), *A compact shape representation for linear geographical objects: the scope histogram*. ACM-GIS, Arlington, Virginia, USA, 51-58p.

Shokoufandeh, A., L. Bretzner, D. Macrini, M.F. Demirci, C. Jonsson and S. Dickinson. (2006), "The representation and matching of categorical shape". *Computer Vision and Image Understanding*, Vol. 103(2): 139-154.

Stallins, J.A. and A.J. Parker. (2003), "The influence of complex systems interactions on barrier island dune vegetation pattern and process". *Annals of the Association of American Geographers*, Vol. 93(1): 13-29.

Turner, M.G. (2005), "Landscape ecology: what is the state of the science?". *Annual Review of Ecology, Evolution, and Systematics*, Vol. 36: 319-344.

Zhang, D.S. and G.J. Lu. (2004), "Review of shape representation and description techniques". *Pattern Recognition*, Vol. 37(1): 1-19.

Zhang, J., X. Zhang, H. Krim and G.G. Walter. (2003), "Object representation and recognition in shape spaces". *Pattern Recognition*, Vol. 36(5): 1143-1154.

# 5

## Projective Spray Can Geometry – An Axiomatic Approach to Error Modelling for Vector Based Geographic Information Systems

*Gwen Wilke & Andrew U. Frank*

## Abstract

*The integration of error analysis in Geographic Information Systems (GIS) is a dominating research topic in Geographic Information Science. Current vector-based GIS software is based on idealized geometric objects: infinitely small points and infinitely thin lines disregard the real character of object representation. The present paper outlines an axiomatic model of 2D projective geometry that incorporates positional random error. As a basis, Menger's axiomatic system for projective geometry is used. Points with errors are modelled by normal distributions and are called 'spray can points'. To model lines with errors an extended version of the duality principle of projective geometry is defined, called 'spray can duality'. Analogous to the exact case, spray can duality is an involutary one-to-one mapping of subsets of the modelled space. Equipped with spray can duality lines with errors or 'spray can lines' can be defined to be the spray can duals of spray can points. The paper gives proof that these objects fulfill the axioms of projective geometry as proposed by Menger and are therefore suitable as a projective model. In future work the proposed model will be used to formulate a corresponding Euclidean model for objects with errors.*

## 5.1 Introduction

A fundamental issue related to Geographic Information Systems (GIS) is the digital representation and manipulation of spatial objects. The basic geometric components for the representation of objects in vector-based GIS are points, lines and polygons. Operations for object manipulation are based on the rules of Euclidean geometry. Geometric functionality in GIS is implemented on the basis of a model of infinitely small points and infinitely thin lines. This is in sharp contrast to the fact that geographic data and their representation are extended and uncertain in location.

The integration of error representation and analysis is often regarded as crucial for the commercial and legal viability of GIS. Users should be able to assess the accuracy of the information upon which they base their decisions. One aspect of error analysis is the assessment of error propagation effects during GIS operations. The propagation of errors may sum up effects and, in the worst case, lead to meaningless data. Consequently, it should be an integral part of GIS operations.

In the present paper an axiomatic model of 2D projective geometry is defined that incorporates random error in positional spatial data. Projective geometry provides a more concise axiomatic system than Euclidean geometry. It can be formulated in simple algebraic terms and independently of dimension. The powerful concept of duality allows exchanging the notions of *point* with *line* and *connect (join)* with *intersect (meet)* and vice versa without violating the validity of a theorem (Coxeter, 1969). As a consequence, an axiomatic system of projective geometry requires only half of the axioms, provided a duality operation has been defined. Projective geometry is preferable as a foundation for geometric operations for GIS. Since Euclidean space can be naturally embedded in the projective space, this approach provides a basis for modelling error propagation in Euclidean geometry.

A point in the proposed model is called a *spray can point*, and is modelled by the two dimensional probability density function (PDF) of a Gaussian normal distribution. Random error in geometric data is usually assumed to be normally distributed (Ghilani and Wolf, 2006). This assumption is based on the central limit theorem stating that the sum of identically independent distributions approximately follows a normal distribution. The name *spray can geometry* is motivated by the way a spray can produces points. The single droplets of paint are randomly distributed over the paper, following a Gaussian distribution. The probability of one droplet falling in the area *dxdy* is given by the integral over the Gaussian PDF in *dxdy*.

Starting with the definition of a *spray can point*, *spray can lines* are introduced by employing a spray can version of duality. Based on the axioms of projective geometry proposed by Menger (Blumenthal and Menger, 1970) *spray can operations* are defined for the connection of spray can points and the intersection of spray can lines. The present paper gives proof that the defined objects and operations fulfill Menger's axioms of projective geometry. The novel contribution is the axiomatic approach to define operations for geometric objects with error.

This paper is divided into five sections. Following the introduction, a brief review of literature related to modelling positional random errors in 2D GIS is given. Subsequent to a definition of axiomatic geometry in Section 5.3, the proposed spray can model of projective geometry is introduced in Section 5.4. Section 5.5 concludes the paper and addresses open questions.

## 5.2 Modelling positional random error

The idea for the proposed research emerged from a keynote speech given by Lotfi Zadeh (2007) about computing with imprecise data. He suggested an approach replacing the ideal model of extensionless points and lines by points and lines as they are produced by a spray can.

Several approaches exist for assessing the positional error of derived geometric objects in vector-based GIS. One common way is the simulation method. A Monte Carlo approach can be applied to simulate the probability density function of a line segment based on the PDF of the endpoints (Lei *et al.*, 2006). Abbaspour *et al.*, (2003) use Monte Carlo simulation for testing the error propagation behaviour of overlay operations for polygons. Heuvelink *et al.* (2007) specify the positional uncertainty behaviour of a geometrical object by assigning appropriate PDFs to primitive points of the object (e.g. the corners of a polygon) or to reference points of the object (e.g. the object's centroid). Realizations of the modelled behaviour can be used as input for Monte Carlo uncertainty propagation studies. In contrast to this, the present paper gives an axiomatic model of geometry that incorporates error propagation as an integral part.

Clementini (2005) specifies a model of uncertain lines as an extension of the model for regions with a broad boundary. In contrast to the present work, the model aims at the study of topological relations between uncertain lines.

Another way of modelling positional error of geometric objects is to use buffer based models: if the accuracy and dependencies of the coordinates of a point is known, the positional uncertainty of a point can be represented by the standard error ellipse. For line segments the concept of a buffer-based model goes back to Perkal's definition of the epsilon band model (cited in Leung *et al.*, 2003). Many variants of the epsilon band model have been developed since then. Alesheikh and Li (1996) represent the error of a line segment by the union of error ellipses of all points along that line depending on the positional errors of the two endpoints. Shi (1998) introduced the error band model with a rectangular confidence region of non-uniform width. He considers the endpoints of the line segment to be statistically independent. This model was generalized by Shi and Liu (2000) for statistically dependent endpoints, and named the G-band model.

Leung *et al.* (2004) introduced the covariance-based error band model for line segments that contains the classical epsilon band model and the error band model as special cases. Assuming a multivariate normal distribution for points with errors, lines with errors are derived by strictly applying the approximate law of error propagation. Formulas are derived for the intersection of lines with errors, resulting again in points with errors. Closedness of operations is achieved. The present approach likewise addresses the issue of closedness of operations, but differs from Leung *et al.* (2004) by aiming at an axiomatic approach. An axiomatic approach has the advantage that it produces consistent solutions for all derived operations and relations, because consistency is guaranteed.

In his thesis, Heuel (2004) proposes a method for statistical reasoning for polyhedral object reconstruction. He proposes a framework of uncertain projective geometry using Grassman-Cayley algebra. His work is similar to the approach presented in this paper, but focuses on photogrammetric applications. The formalism used builds on algebraic invariants of projective geometry, but is not a direct model of an axiomatic system.

The research proposed here aims at making use of the power of a closed and fully defined axiomatic system. An error model that complies with the axioms of projective geometry automatically provides consistency, i.e., it avoids contradictions. Equipped with a realistic interpretation of geometric primitives, all geometric objects can be derived. The novel contribution of this work is the use of axiomatic geometry to build an error model of objects under positional random error.

## 5.3   Axiomatic Projective Geometry

### 5.3.1   Coexisting models of geometry

Euclid's *Elements* is often referred to as the most successful textbook in the history of mathematics, since it introduced the deductive method to formal sciences. Its aspiration was to logically deduce all theorems of geometry from few obvious statements and rules that need not be justified. These statements are nowadays called *undefined terms* or *geometric primitives*; the rules are called *postulates* or *axioms*.

A *model* of Euclidean geometry is an interpretation of primitives that fulfills the axioms. A metaphor commonly attributed to the German mathematician David Hilbert states that it should always be possible to replace the notions *point, line* and *plane* by *beer mug, bank* and *desk*, as long as the rules apply. The modern and fully consistent axiomatic approach determines the classical Euclidean geometry up to isomorphism.

The principle of coexisting isomorphic models can be employed for all modern geometric systems, including projective geometry. It is used in the present paper to define a *spray can model* of projective geometry. *Spray can objects* are interpretations of primitive objects incorporating positional random error. *Spray can relations* are interpretations of primitive relations that operate on spray can objects. They must be defined accordingly, so that the axioms apply.

### 5.3.2 Menger's axiomatization of projective geometry

Menger's axiomatization is a purely algebraic approach which, as a consequence, is dimension independent. The primitive objects of Menger's axiomatic system are a set $S$ of objects of concern and two special objects contained in $S$. These special objects are *vacuum* $V$, representing *nothing*, and *universe* $U$, representing *everything*. The primitive operations in Menger's definition are *join* and *meet*. *Join*, denoted by $\vee$, connects two elements of $S$; *meet*, denoted by $\wedge$, intersects two elements of $S$.

The role of *vacuum* and universe is *axiomatized* by four postulates:

$$X \vee V = X, \quad X \wedge V = V \tag{5.1}$$

$$X \wedge U = X, \quad X \vee U = U \tag{5.2}$$

for all $X$ in $S$. The behaviour of arbitrary projective objects is determined by two axioms called *projective laws*:

$$X \vee ((X \vee Y) \wedge Z) = X \vee ((X \vee Z) \wedge Y) \tag{5.3}$$

$$X \wedge ((X \wedge Y) \vee Z) = X \wedge ((X \wedge Z) \vee Y) \tag{5.4}$$

The axioms of Equation 5.1 and Equation 5.2 are dual to each other: Equation 5.2 is obtained from Equation 5.1 by replacing each occurrence of $U$ by $V$ and each occurrence of $\vee$ by $\wedge$ and vice versa. The same holds for the axioms of Equation 5.3 and Equation 5.4.

Menger's axiomatization is a dimension-independent definition of projective geometry. In two-dimensional projective geometry, the set $S$ of objects of concern is called the *projective plane* $P^2$. It resolves into four distinct groups: $U$, $V$, points and lines. $U$ is dual to $V$ and points are dual to lines. The *duality principle of projective plane geometry* states, that a valid theorem remains true, if the following terms are interchanged:

$$
\begin{array}{ccc}
\text{point} & \leftrightarrow & \text{line} \\
V & \leftrightarrow & U \\
\text{join} & \leftrightarrow & \text{meet}
\end{array}
$$

To define a model of projective geometry, it is sufficient to define instances of the primitives *point*, $V$ and *join*, and a *duality* operation for them.

### 5.3.3 The spherical model of the real projective plane

A common way to model the two-dimensional real projective plane $P^2$ is the unit sphere $S^2 \subset R^3$, where antipodal points are identified (Figure 5.1a). In this model projective points are represented by antipodal pairs of $R^3$-points on the sphere, and projective lines are represented by great circles. More precisely, a projective point $p$ can be represented by a pair of $R^3$-points $(\bar{p},-\bar{p})$, where $\bar{p}$ has unit length. Each pair of points uniquely determines a line through the origin in $R^3$. A projective line $l \in P^2$ can be represented by a great circle in $S^2$. Each great circle uniquely determines a plane through the origin of $R^3$.

Projective points and lines are *dual* to each other. The duality operation $d : P^2 \to d(P^2)$ uniquely maps each point $p$ on the sphere to the great circle perpendicular to $p$ (Figure 5.1a): $d(p) := p^{\perp} \cap S^2$, where $p^{\perp}$ is the orthogonal complement of $p$ in $R^3$. The operation d is an involution, i.e. it is a function that is its own inverse:

$$d.d = id \tag{5.5}$$

where . denotes the composition of functions and *id* denotes the identity operation. Consequently projective lines can be represented by pairs of $R^3$-points in the dual space.

The vacuum $V = \{\vec{0}\}$ is represented by the singleton containing the origin $\vec{0} \in R^3$. It is dual to the universe $U = V^{\perp} \cap S^2 = R^3 \cap S^2 = S^2$.

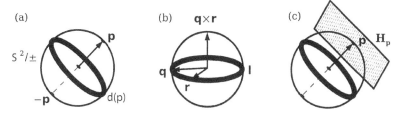

**Figure 5.1.** (a) Duality in the unit sphere model of the projective plane, (b) two projective points $q,r$ determine a projective line $l = d(q \times r)$, (c) the homogeneous plane $H_p$.

Two projective objects can be connected by a *join* operation:

$$x \vee y := (x + y) \cap S^2 \tag{5.6}$$

where + denotes the sum of linear subspaces in $R^3$. If $q$ and $r$ are distinct points their connection can be calculated using the cross product:

$$q \vee r = d(q \times r) \quad \in P^2 \tag{5.7}$$

where $(q \times r) \in d(P^2)$ is a projective point in the dual space and $d(q \times r) := l \in P^2$ is the line dual to $(q \times r)$ in the primal space (Figure 5.1b). Dually, two projective objects can be intersected by a *meet* operation:

$$x \wedge y := x \cap y = \left(x^{\perp} + y^{\perp}\right)^{\perp} = d(d(x) \vee d(y)) \tag{5.8}$$

where $\cap$ denotes the intersection of linear subspaces in $R^3$. If $l$ and $m$ are distinct lines, their intersection can be calculated using the cross product: $l \wedge m = d\big(d(l) \times d(m)\big)$, where $d(l)$ and $d(m)$ are points in the dual space.

The spherical model of projective geometry satisfies Menger's Axioms (Blumenthal and Menger, 1970). Note that for linear subspaces $x + (y \cap z) \neq (x + y) \cap (x + z)$.

## 5.4   The Spray Can Model for the Real Projective Plane

### 5.4.1   Spray can points and spray can vacuum

The PDF of the two-dimensional *Gaussian normal distribution* $G_{(\mu,\Sigma)}(x) : R^2 \rightarrow R^2$ is defined on the Euclidean plane $R^2$. $G$ is uniquely determined by its mean $\mu$ and its covariance matrix $\Sigma$:

$$G_{(\mu,\Sigma)}(x) = \left(2\pi \cdot \sqrt{|\Sigma|}\right)^{-1} \cdot \exp\left(-\frac{1}{2}(x-\mu)^T \Sigma^{-1}(x-\mu)\right). \tag{5.9}$$

For a point $p \in S^2$ let $B_p = \{b_1, b_2, p\}$ be an orthonormal basis of $R^3$ and $B'_p = \{b_1, b_2\}$ its two-dimensional restriction to the orthogonal subspace perpendicular to $p$. The transformation of a point $x = [x_1, x_2]_{B'_p} \in R^2$ to its homogeneous coordinates with respect to $B_p$ is given by:

$$h_{B_p} : R^2 \rightarrow H_p, \quad [x_1, x_2]_{B'_p} \mapsto [x_1, x_2, 1]_{B_p} \tag{5.10}$$

where $H_p$ is the homogeneous plane with respect to $B_p$ (Figure 5.1c),

$$H_p = \left(\left\{ x = [x_1, x_2, x_3]_{B_p} : x_3 \neq 0 \right\} / \equiv\right) \quad \text{where} \quad x \equiv \bar{x} \Leftrightarrow x = k \cdot \bar{x}, k \in R, k \neq 0. \tag{5.11}$$

The function $h_{B_p}$ is an isomorphism, i.e. $H_p \cong R^2$. The inverse transformation is given by

$$h_{B_p}^{-1} : H_p \rightarrow R^2, \quad [x_1, x_2, x_3]_{B_p} \mapsto [x_1/x_3 , x_2/x_3]_{B'_p}. \tag{5.12}$$

For $O = (0,0) \in R^2$ we have $h_{B_p}(O) = [0,0,1]_{B_p} = p$ and hence $h_{B_p}^{-1}(p) = O$.

**Definition 1 (Spray can point).** For a given exact projective point $p \in P^2$ and a covariance matrix $\Sigma \in R^4$ a *spray can point (scp)* $\tilde{p}_\Sigma$ is a function $\tilde{p}_\Sigma : P^2 \rightarrow R$:

$$\tilde{p}_\Sigma(x) = \begin{cases} G_{(O,\Sigma)}\big(h_{B_p}^{-1}(x)\big) & \text{if } x \in H_p \\ 0 & \text{else}, \end{cases} \tag{5.13}$$

where $G(O,\Sigma)$ is the PDF of a Gaussian normal distribution on $H_p \cong R^2$ with mean $\mu = O = (0, 0) \in R^2$ and covariance matrix $\Sigma$.

A *scp* assigns every exact projective point $x \in P^2$ the corresponding value of the Gaussian PDF (Figure 5.2a). We call the point $p$ *base point* of $\tilde{p}_\Sigma$. It is assigned the unique maximum value of the Gaussian PDF.

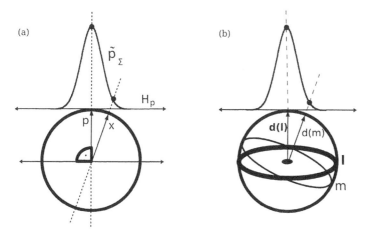

**Figure 5.2.** (a) A spray can point $\tilde{p}$ in $p$ with covariance matrix $\Sigma$ assigns a value to every projective point $x$, (b) A spray can line $\tilde{l}_\Sigma$ in $l$ assigns a value to every projective line $m$.

**Definition 2 (Spray can vacuum).** The constant map $\tilde{V}_{\Sigma_V} : V = \{0\} \mapsto 0$ is called *spray can vacuum*. Since $\tilde{V}_{\Sigma_V}$ does not depend on the covariance matrix $\Sigma_V$, we can consider $\Sigma_V$ arbitrary.

### 5.4.2  Spray can duality

A spray can version of duality can be defined by utilizing the duality operation $d : P^2 \rightarrow d(P^2)$ on the exact spherical model of the projective plane (Figure 5.1a). $d$ maps an exact projective object $x$ in the primal space $P^2$ to an exact projective object $dx$ in the dual space $dP^2$. Likewise, the *spray can duality* operation $\tilde{d}$ maps a spray can object $\tilde{x}$ to its spray can dual object $\tilde{dx}$.

**Definition 3 (Spray can dual).** We define the *spray can dual (scd)* of a spray can point $\tilde{p}_\Sigma$ and the spray can vacuum $\tilde{V}_\Sigma$ by:

$$\tilde{d}.\tilde{p}_\Sigma = \tilde{p}_\Sigma.d \quad \text{and} \quad \tilde{d}.\tilde{V}_\Sigma = \tilde{V}_\Sigma.d \tag{5.14}$$

respectively.

In other words $\tilde{dp}_\Sigma(x) = \tilde{p}_\Sigma(dx)$ for all $x \in dP^2$, i.e. $\tilde{dp}_\Sigma : dP^2 \rightarrow R$ is a spray can point in the dual space. Likewise, the spray can dual of the spray can vacuum is the spray can vacuum of the dual space. Consequently $\tilde{d}.\tilde{d}.\tilde{p}_\Sigma$ and $\tilde{d}.\tilde{d}.\tilde{V}_\Sigma$ are well defined. Since $d$ is an involution (cf. Equation 5.5) $\tilde{d}$ is involutary as well: $\tilde{d}.\tilde{d}.\tilde{x}_\Sigma = \tilde{x}_\Sigma.d.d = \tilde{x}_\Sigma.id = \tilde{x}_\Sigma$ for $x = p$ or $x = V$. We will see in chapter 4.4 that $\tilde{dp}_\Sigma$ can be interpreted as a spray can line and that $\tilde{dV}_\Sigma$ can be interpreted as the spray can universe. $\tilde{dx}_\Sigma$ inherits the covariance matrix $\Sigma$ from its primal spray can object $\tilde{x}_\Sigma$. Since $\tilde{dx}_\Sigma(dx) = \tilde{x}_\Sigma(x)$ holds $\tilde{dx}_\Sigma$ has base object $dx$. It is convenient to abbreviate $\tilde{dx}_\Sigma =: (dx)_\Sigma^\sim$.

**Definition 4 (Spray can object).** A *spray can object (sco)* is any real valued map $\tilde{x}_\Sigma$ defined on $P^2$ or $dP^2$ that can be generated by applying the spray can dual $\tilde{d}$ to a spray can point or to the spray can vacuum. $\tilde{x}_\Sigma$ depends on the covariance matrix $\Sigma \in R^4$. $x \in P^2$ is called *base object* of $\tilde{x}_\Sigma$.

As a consequence of the involutary property of the spray can dual operation four types of spray can objects are differentiated: spray can points, their spray can duals, the spray can vacuum and its spray can dual.

### 5.4.3 Spray can join

A spray can version of the *join* of two spray can objects can be defined by using the join operator $\vee$ of the exact spherical model (cf. Equations 5.6 and 5.7). In the following we will use the symbol $\vee$ for both models. Since all types of spray can objects can be generated from spray can points and the spray can vacuum by spray can duality, we may assume that every initial data set consists only of point measurements. As a consequence of this assumption it is necessary to define an operational combination of all different types of spray can objects generated from spray can points and spray can vacuum.

**Definition 5 (Spray can join).** For spray can objects $\tilde{x}, \tilde{y}$ with covariance matrices $\Sigma_x, \Sigma_y$ a *Spray can join (scjoin)* operation $\vee$ is defined by:

$$\tilde{x}_{\Sigma_x} \vee \tilde{y}_{\Sigma_y} := (x \vee y)_{\tilde{\Sigma}_{x,y}} ,\qquad (5.15)$$

where $\Sigma_{x,y}$ can be defined assuming that every initial data set consists only of point measurements: Given spray can points $(\tilde{p}_1)_{\Sigma_{p1}},(\tilde{p}_2)_{\Sigma_{p2}},(\tilde{p}_3)_{\Sigma_{p3}},(\tilde{p}_4)_{\Sigma_{p4}}$, with base points $p_1, p_2, p_3, p_4$ and covariance matrices $\Sigma_{p1}, \Sigma_{p2}, \Sigma_{p3}, \Sigma_{p4}$ respectively, the covariance matrix $\Sigma_{x,y}$ of their connections is given by:

$$\Sigma_{V,p_1} = \Sigma_{p_1,V} = \Sigma_{dV,p_1} = \Sigma_{p_1,dV} := \Sigma_{p_1} ,$$

$$\Sigma_{p_1,p_2} := \begin{cases} \Sigma_{p_1 \times p_2} & \text{if } p_1 \neq p_2 \\ choose(\Sigma_{p_1}, \Sigma_{p_2}) & \text{if } p_1 = p_2 \end{cases} ,$$

$$\Sigma_{p_1,(p_2,p_3)} := \begin{cases} \Sigma_{p_1} & \text{if } p_1 \neq p_2 \neq p_3 \ \text{not coplanar} \\ choose(\Sigma_{p_i \times p_j})_{\substack{i \neq j, \\ i,j \in \{1,2,3\}}} & \text{if } p_1 \neq p_2 \neq p_3 \ \text{coplanar} \\ choose(\Sigma_{p_1 \times p_3}, \Sigma_{p_2 \times p_3}) & \text{if } p_1 = p_2 \neq p_3 \ \text{coplanar} \end{cases} , \qquad (5.16)$$

where $\Sigma_{p_1 \times p_2}$ is obtained by applying the approximate law of error propagation to $\Sigma_{p_1}, \Sigma_{p_2}$ over the cross product. The function *choose* chooses the covariance matrix that was obtained from a minimum number of construction steps, i.e., it chooses the presumably most accurate measurement of $\tilde{p}_{\Sigma}$. If this criterion is not decisive, i.e., in the case where two spray can points $\tilde{p}_{\Sigma}$ and $\tilde{p}_{\Pi}$ with $\Sigma \neq \Pi$, result from an equal number of construction steps or, in case of two measurements of the same point with different accuracy, an expert has to decide which representation to choose. As long as the expert's choice does not change, no contradictions occur. From Definition 3 $\Sigma_V = \Sigma_{dV}$ and $\Sigma_{p \times q} = \Sigma_{d(p \times q)}$ follows. Note that $\Sigma_{dV} = \Sigma_V$ holds formally, whereas in fact both matrices can be chosen arbitrarily.

In practice the interesting instance of Definition 5 is the join of two spray can points with different base points: For spray can points $\tilde{p}_{\Sigma_p}, \tilde{q}_{\Sigma_q}$, $p \neq q$, Definition 5 yields:

$$[\tilde{p}_{\Sigma_p} \vee \tilde{q}_{\Sigma_q}](x) = [(p \vee q)]_{\Sigma_{p,q}}^{\sim}(x) = \tilde{d}.\tilde{d}.[(p \vee q)]_{\Sigma_{p,q}}^{\sim}(x) = \tilde{d}.[d(p \vee q)]_{\Sigma_{p,q}}^{\sim}(x)$$

$$= \tilde{d}.(p \times q)_{\Sigma_{p \times q}}^{\sim}(x) = (p \times q)_{\Sigma_{p \times q}}^{\sim}(dx) \quad \forall x \in P^2 \qquad (5.17)$$

where $(p \times q)\widetilde{\Sigma}_{p \cdot q}$ is a *scp* in the dual space (cf. Equation 5.7) with base point $p \times q$. We will see in the following chapter that $\tilde{p}_{\Sigma_p} \vee \tilde{q}_{\Sigma_q}$ can be interpreted as a spray can line.

In the case of two spray can points with identical base points, the expert has to choose the initial *scp* she wants to base her subsequent constructions on. In the case of three spray can points with different but coplanar base points, the expert has to choose an initial pair of spray can points to construct a spray can line.

### 5.4.4 Construction of dual primitives

Provided with the definitions of *scp, spray can vacuum* and *scjoin* the concepts of *spray can line, spray can meet* and *spray can universe* follow by *spray can duality* $\tilde{d}$.

**Definition 6 (Spray can line).** We define a *spray can line (scl)* $\tilde{l}_{\Sigma} : P^2 \to P^2$ with base line $l = dp \in P^2$ and covariance matrix $\Sigma \in R^4$ by:

$$\tilde{l}_{\Sigma} := \tilde{d}\tilde{p}_{\Sigma} \tag{5.18}$$

where $\tilde{p}_{\Sigma}$ is a *scp* in $dP^2$ with base point $p$.

$\tilde{l}_{\Sigma}$ is the spray can dual of a given *scp* $\tilde{p}_{\Sigma}$ in the dual space $dP^2$ with base point $p$. With the abbreviation $\tilde{dp}_{\Sigma} := (dp)_{\widetilde{\Sigma}}$ for spray can points, Definition 6 reads $\tilde{l}_{\Sigma} = (dp)_{\widetilde{\Sigma}}$. $\tilde{l}_{\Sigma}$ assigns every exact projective line $m$ the value of its dual point $dm$ under $\tilde{p}_{\Sigma}$ (Figure 5.2b). Since spray can duality is self-inverse the spray can dual of a *scl* $\tilde{l}_{\Sigma}$ is a *scp*, $\tilde{d} \cdot \tilde{l}_{\Sigma} = \tilde{d} \cdot \tilde{d} \cdot \tilde{p}_{\Sigma} = \tilde{p}_{\Sigma}$. As a consequence of Definitions 5 and 6 the result of the *scjoin* operation of two spray can points is a spray can line (cf. Equation 5.17).

**Definition 7 (Spray can meet).** We define a *spray can meet (scmeet)* operation $\wedge$ for the intersection of two spray can objects $\tilde{x}_{\Sigma_x}$ and $\tilde{y}_{\Sigma_y}$ analogous to the exact meet operation by:

$$\tilde{x}_{\Sigma_x} \wedge \tilde{y}_{\Sigma_y} := \tilde{d}\left( \tilde{d}\tilde{x}_{\Sigma_x} \vee \tilde{d}\tilde{y}_{\Sigma_y} \right). \tag{5.19}$$

Because of:

$$\tilde{d} \cdot \left( \tilde{d}\tilde{x}_{\Sigma_x} \vee \tilde{d}\tilde{y}_{\Sigma_y} \right) = \tilde{d} \cdot \left[ (dx)_{\widetilde{\Sigma}_x} \vee (dy)_{\widetilde{\Sigma}_y} \right] =$$
$$= \tilde{d} \cdot \left[ (dx \vee dy)_{\widetilde{\Sigma}_{x,y}} \right] = \left[ d(dx \vee dy) \right]_{\widetilde{\Sigma}_{x,y}} = [x \wedge y]_{\widetilde{\Sigma}_{x,y}} \tag{5.20}$$

we can write $\tilde{x}_{\Sigma_x} \wedge \tilde{y}_{\Sigma_y} = (x \wedge y)\tilde{\Sigma}_{x,y}$ for short. Note that $\tilde{x}_{\Sigma_x} \wedge \tilde{y}_{\Sigma_y}$ and $\tilde{x}_{\Sigma_x} \vee \tilde{y}_{\Sigma_y}$ have the same covariance matrix $\Sigma_{x,y}$ attached.

The spray can meet of $n > 2$ spray can objects is well defined by the pertinent instances of the spray can join operation (Definition 5). The mixed case of *scmeet* and *scjoin* operations is well defined by combining Definitions 3, 5 and 7. To see this it is sufficient to check a mixed combination of operations for three spray can objects $\tilde{x}_{\Sigma_x}, \tilde{z}_{\Sigma_z}$ and $\tilde{y}_{\Sigma_y}$ :

$$\left( \tilde{x}_{\Sigma_x} \vee \tilde{y}_{\Sigma_y} \right) \wedge z_{\Sigma_z} = (x \vee y)_{\widetilde{\Sigma}_{x,y}} \wedge z_{\Sigma_z} = \left[ (x \vee y) \wedge z \right]_{\widetilde{\Sigma}_{(x,y),z}}. \tag{5.21}$$

**Definition 8 (Spray can universe).** $\tilde{U}_{\Sigma_U} := \tilde{d} \cdot \tilde{V}_{\Sigma_V}$ is called the *spray can universe*. Note that the spray can universe is the constant map

$\bar{U}_{\Sigma_U} = \tilde{V}_{\Sigma_V} . d : U = \{0\}^{\perp} = R^3 \mapsto 0$. It does not depend on the covariance matrix $\Sigma_U$, which is considered arbitrary and we can write $\Sigma_U = \Sigma_V$.

Spray can duality maps spray can objects in the primal space $P^2$ to spray can objects in the dual space $d(P^2)$ in the following way:

$$\begin{aligned} scp &\leftrightarrow scl \\ \tilde{V} &\leftrightarrow \tilde{U} \\ scjoin &\leftrightarrow scmeet \end{aligned} \tag{5.22}$$

### 5.4.5  Verification of the axioms

In this subsection, Menger's axioms of projective geometry (Equations 5.1-5.4) are verified for the objects and operations of the spray can model introduced in the foregoing chapters. As main tools the duality principle Equation 5.22 for spray can objects and Menger's axioms for exact projective objects are applied.

**Proof of Equations 5.1 and 5.2.** The first axiom of Equation 5.1 holds by Definition 5: $\tilde{x}_{\Sigma_x} \vee \tilde{V}_{\Sigma_V} = (x \vee V)_{\tilde{\Sigma}_x} = \tilde{x}_{\Sigma_x}$. The second axiom of Equation 5.1 follows by Definition 7: $\tilde{x}_{\Sigma_x} \wedge \tilde{V}_{\Sigma_V} = (x \wedge V)_{\tilde{\Sigma}_{x,y}} = \tilde{V}_{\Sigma_x} = \tilde{V}_{\Sigma_V}$. The two axioms of Equation 5.2 follow by spray can duality: $\tilde{x}_{\Sigma_x} \wedge \tilde{U}_{\Sigma_U} = (x \wedge U)_{\tilde{\Sigma}_{x,y}} = \tilde{x}_{\Sigma_x}$ and $\tilde{x}_{\Sigma_x} \vee \tilde{U}_{\Sigma_U} = (x \vee U)_{\tilde{\Sigma}_{x,y}} = \tilde{U}_{\Sigma_x} = \tilde{U}_{\Sigma_U}$.

**Proof of Equations 5.3 and 5.4.** To verify the axiom represented by Equation 5.3, note that for the indices in Equation 5.16 commutativity holds, because both, the function *choose* and the cross product modulo opposite points, are commutative. Associativity of indices holds for three arguments. This can be seen as follows: For the case of three different base points $p_1 \neq p_2 \neq p_3$ that are not coplanar $(p_1 \vee p_2) \vee p_3 = p_1 \vee (p_2 \vee p_3) = U$ holds and the associated covariance matrix can be chosen arbitrarily. For coplanar base points with $p_1 \neq p_2 \neq p_3$ associativity follows from the associativity of *choose*. For coplanar base points with $p_1 = p_2 \neq p_3$ $\Sigma_{(p_1,p_2)} = Choose(\Sigma_{p_1}, \Sigma_{p_2})$ holds and $\Sigma_{(p_1,p_2),p_3} = Choose(\Sigma_{p_1,p_3}, \Sigma_{p_2,p_3}) = \Sigma_{p_1,(p_2,p_3)}$. Applying Definition 5 to the left hand side of Equation 5.3 yields:

$$\begin{aligned} \tilde{x}_{\Sigma_x} \vee \left((\tilde{x}_{\Sigma_x} \vee \tilde{y}_{\Sigma_y}) \wedge \tilde{z}_{\Sigma_z}\right) &= \tilde{x}_{\Sigma_x} \vee \left[(x \vee y) \wedge z\right]_{\tilde{\Sigma}_{(x,y),z}} \\ &= \left[x \vee \left((x \vee y) \wedge z\right)\right]_{\tilde{\Sigma}_{x,((x,y),z)}} \\ &\overset{(*)}{=} \left[x \vee \left((x \vee z) \wedge y\right)\right]_{\tilde{\Sigma}_{x,((x,z),y)}} \\ &= \tilde{x}_{\Sigma_x} \vee \left((\tilde{x}_{\Sigma_x} \vee \tilde{z}_{\Sigma_z}) \wedge \tilde{y}_{\Sigma_y}\right). \end{aligned} \tag{5.23}$$

In step (*) the axiom represented by Equation 5.3 applies for exact projective objects and $\Sigma x,((x,y),z) = \Sigma x,((x,z),y)$ follows from commutativity and associativity of indices by $(x,y), z = x,(y,z) = x,(z,y) = (x,z), y$ (cf. Definition 5). This proves axiom Equation 5.3 for spray can objects. Equation 5.4 holds by spray can duality.

## 5.5  Conclusions and Further Work

We have shown that an axiomatization of a realistic treatment of projective geometry can be achieved. The long term objective of this research is to define a pertinent realistic model of Euclidean geometry for GIS. Current GIS are based on an idealized axiomatization of geometry which deals with points and lines without extensions. Real points and lines have extensions and the locations of idealizations are not precisely measurable. Lotfi Zadeh (2007) has suggested the geometry produced when drawing with a spray can as an inspiration for a realistic model of geometry.

A spray can produce a point with a random distribution of color droplets, approximating a Gaussian normal distribution. The paper drafts how to construct a consistent axiomatic geometry with this model.

For an axiomatization we use Menger's axioms of projective geometry, which are fewer and simpler than Hilbert's axioms for Euclidean geometry (Blumenthal and Menger, 1970; Hilbert, 1968). The duality between points and lines in projective geometry further simplifies the task. It is necessary to show how to embed a spray can point into the projective plane and then define two of the three operations *intersection*, *connection* and *duality*. We have given definitions for *duality* and *connection*. *Intersection* follows by duality.

The used axiomatic approach assures consistency for the treatment of all geometric operations derived from this foundation. The advantage of these definitions over other more pragmatic efforts in the past is that the implementation of the basic operations is sufficient to extend derived operations, analytical functions and tests for relations to treat geometry with positional uncertainty without danger of contradiction.

In a next step the *join* and *meet* operations introduced in this paper will be extended to allow a dimension independent formulation of the model. 2D and 3D operations will be formulated as special cases. The definitions introduced must be extended and modified to define a model of Euclidean geometry with uncertainty in the location. This model will allow the integration of error analysis and error propagation in current GIS software by accessing standard operations of vector-based GIS.

# References

Abbaspour, R.A., M. Delavar and R. Batouli. (2003), "The issue of uncertainty propagation in spatial decision making". *In:* K. Virrantaus and H. Tveite (eds.) *ScanGIS'2003. The 9th Scandinavian Research Conference on Geographical Information Science, 4-6 June 2003, Espoo, Finland*, Department of Surveying, Helsinki University of Technology, pp. 57-65.

Alesheikh, A.A. and R. Li. (1996), "Rigorous uncertainty models of line and polygon objects in GIS". *In:* American Congress on Surveying and Mapping. *GIS/LIS '96 Annual Conference and Exposition Proceedings,* American Society for Photogrammetry and Remote Sensing, Bethesda, Md, USA, pp. 906-920.

Blumenthal, L. and K. Menger. (1970), *Studies in Geometry*, W.H. Freeman and Company, San Francisco, 512p.

Clementini, E. (2005), "A model for uncertain lines". *Journal of Visual Languages & Computing*, Vol. 16: 271-288.

Coxeter, H.S.M. (1969), *Introduction to Geometry* (2nd edn.), Wiley and Sons, New York, 469p.

Ghilani, C.D. and P.R. Wolf. (2006), *Adjustment Computation: Spatial Data Analysis* (4<sup>th</sup> edn.), John Wiley and Sons, Hoboken, New Jersey, 640p.

Heuel, S. (2004), *Uncertain Projective Geometry*. PhD thesis, Institute for Photogrammetry, Rheinische Friedrich-Wilhelms-Universität, Bonn, Germany, 205p.

Heuvelink, G., J. Brown and E. van Loon. (2007), "A probabilistic framework for representing and simulating uncertain environmental variables". *International Journal of Geographical Information Science*, Vol. 21(5): 497-513.

Hilbert, D. (1968), *Grundlagen der Geometrie*. Teubner Studienbücher Mathematik, Stuttgart, 271p.

Lei, Z., D. Min and C. Xiaoyon. (2006), "A new approach to simulate positional error of line segment in GIS". *Geo-Spatial Information Science*, Vol. 9(2): 142-146.

Leung, Y., J.-H. Ma and M.F. Goodchild. (2003), "A general framework for error analysis in measurement-based GIS – A summary". *In:* W. Shi, M.F. Goodchild and P.F. Fisher (eds.) *Proceedings of the Second International Symposium on Spatial Data Handling*. Hong Kong Polytechnic University, Hong Kong, pp. 23-33.

Leung, Y., J.-H. Ma and M.F. Goodchild. (2004), "A general framework for error analysis in measurement-based GIS Part 1: The basic measurement-error model and related concepts". *Journal of Geographical Systems*, Vol. 6(4): 325-354.

Shi, W. (1998), "A generic statistical approach for modeling error of geometric features in GIS". *International Journal of Geographical Information Science*, Vol. 12(2): 131-143.

Shi, W. and W. Liu. (2000), "A stochastic process-based model for the positional error of line segments in GIS". *International Journal of Geographical Information Science*, Vol. 14(1): 51-66.

Zadeh, L.A. (2007), "Granular computing – computing with uncertain, imprecise and partially true data". *In:* M. Molenaar, W. Kainz and A. Stein (eds.) *Modelling Qualities in Space and Time. Proceedings of the International Symposium on Spatial Data Quality '07 (ISSDQ 2007), ITC, Enschede, The Netherlands*
http://www.itc.nl/ISSDQ2007/proceedings/index.html.

# 6

## Short Notes

### Qualitative Topological Relationships for Objects with Possibly Vague Shapes: Implications on the Specification of Topological Integrity Constraints

*Lotfi Bejaoui, Yvan Bédard, François Pinet, & Michel Schneider*

In spatial databases, natural phenomena are generally represented using geometries with well-defined boundaries. Such a description of reality ignores the shape vagueness that characterizes some spatial objects such as flooding zones, lakes or forest stands. The spatial data quality is therefore degraded because there is a gap between the spatial reality and its description.

Most existing approaches dealing with spatial objects with vague shapes do not allow the representation of partial shape vagueness that concerns recurrent spatial objects, such as a lake with rocky banks on one side and swamp banks on the other side. This poster addresses the problem of representing spatial objects with different levels (*partial, complete*) of shape vagueness. This research proposes a new approach to represent such objects, to compute their topological relationships, and to express the topological constraints involving them. The proposed model is called the *QMM model* (Qualitative Min-Max model). It defines an object with a vague shape as composed of a minimal extent (the part of the object that is certain) and a maximal extent (the part of the object that is uncertain in addition to the minimal extent). The difference between the latter extents refers to the uncertain part of a given object (e.g., the broad boundary; Bejaoui et al., in press). This technique was originally proposed in the Egg-Yolk theory to deal with regions with completely broad boundaries (Cohn and Gotts, 1996a and 1996b). Clementini and Di Felice (1997) proposed a similar model to represent regions and lines with completely broad boundaries, and to identify their topological relationships. In this research, we extend this technique to deal with regions with possibly (partially/completely) broad boundaries, as well as with lines and points with possibly vague shapes. Moreover, a set of adverbs are suggested to describe the shape vagueness of an object (e.g., a region with a *partially* broad boundary, a line with a *completely* broad interior, a line with a partially broad boundary). In the same way, a set of adverbs is proposed to evaluate the uncertainty of a topological relationship between two objects with vague shapes (e.g., *weakly* Contains, *fairly* Contains, *strongly* Covers). Figure 6.1 illustrates our approach to identify topological relationships between regions with broad boundaries. The relationships presented in the figure belong to different levels of *CONTAINS* and *DISJOINT* clusters, respectively, according to the contents of their related matrices.

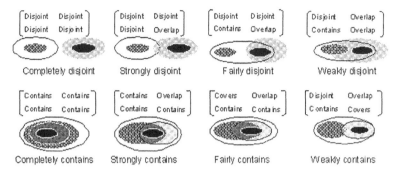

**Figure 6.1.** Examples of topological relationships between regions with broad boundaries according to the QMM model.

This approach is then used to specify the topological integrity constraints involving objects with vague shapes based on the concepts of the QMM model. The topological constraints are expressed using a spatial extension of the Object Constraint Language (OCL; Pinet et *al.*, 2007). Existing software (the *OCL2SQL* tool developed at the University of Dresden; Demuth, 2005) is extended to automatically generate a SQL query from a topological constraint involving objects with vague shapes. This research shows also how the proposed approach can be used to deal with geometries with vague shapes resulting from a vertical integration (the same objects are represented in different source databases using heterogeneous and redundant crisp geometries captured in the same epoch). Figure 6.2 illustrates the problem of vagueness of topological relationships in the spatial data integration process. In Figure 6.2, two spatial objects are represented differently in two data sources *A* and *B*. Three topological relationships are possible between the final geometries resulting from the integration: *Overlap*, *Meet* and *Disjoint*. Then, the topological inconsistencies in the final database are increased. We deal with this problem using the QMM model to specify the topological relationships between final geometries with vague shapes resulting from the integration.

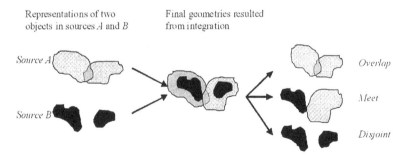

**Figure 6.2.** Possible topological relationships between final geometries with vague shapes.

In conclusion, this approach provides a coarse evaluation of shape vagueness comparative to fuzzy and probabilistic approaches. It does not require a complex data acquisition process. Finally, it is relatively simple to be implemented using existing database management systems.

# References

Bejaoui, L., Y. Bédard, F. Pinet and M. Schneider. (In press), "Qualified topological relations between spatial objects with possibly vague shape". *International Journal of Geographical Information Sciences*.

Clementini, E. and P. Di Felice. (1997), "Approximate topological relations". *International Journal of Approximate Reasoning*, Vol. 16:173-204.

Cohn, A.G. and N.M. Gotts. (1996a), "Representing spatial vagueness: A mereological approach". *In:* L. Carlucci Aiello, J. Doyle, and S. Shapiro (eds.), *KR'96: Principles of Knowledge Representation and Reasoning*. Morgan Kaufmann, San Francisco, pp. 230-241.

Cohn, A.G. and N.M. Gotts. (1996b), "The 'egg-yolk' representation of regions with indeterminate boundaries". *In:* P. Burrough and A.M. Frank (eds.) *Proceedings of the GISDATA Specialist Meeting on Spatial Objects with Undetermined Boundaries*, Taylor and Francis, pp. 171-187.

Demuth B. (2005), "The Dresden OCL Toolkit and the Business Rules Approach". *European Business Rules Conference* (EBRC2005), Amsterdam, http://st.inf.tu-dresden.de/files/papers/EBRC2005_Demuth.pdf, last accessed April 28 2009.

Pinet, F., M. Duboisset and V. Soulignac. (2007), "Using UML and OCL to maintain the consistency of spatial data in environmental information systems". *Environmental modeling & software*, Vol. 22(8):1217-1220.

# A Framework for the Identification of Integrity Constraints in Spatial Datacubes

*Mehrdad Salehi, Yvan Bédard, Mir Abolfazl Mostafavi & Jean Brodeur*

Spatial datacubes extend the datacube concept underlying the field of Business Intelligence (BI) into the realm of spatial analysis, geographic knowledge discovery, and spatial decision-support. Datacubes rely on the multidimensional paradigm defined in the field of statistical analysis. Spatial datacubes provide capabilities not inherent to the transaction-oriented systems such as geographical information systems (GIS) and spatial database engines (universal servers). Consequently, spatial datacubes are central to Spatial On-Line Analytical Processing (SOLAP) and aim at supporting interactive complex analysis involving spatial and temporal data (Bédard *et al.*, 2007).

Data quality in spatial datacubes is an important topic since these databases are used as a basis for decision-making in large organizations dealing with health care, environmental management, and transportation, among others. Indeed, poor quality data in these datacubes may lead to poor decisions that, ultimately, may have significant social and economical impacts. Integrity constraints play a key role in improving logical consistency, i.e., one of the major elements of data quality. More precisely, integrity constraints are assertions typically defined in the conceptual model of an application in order to prevent insertions of incorrect data into a database (Godfrey *et al.*, 1998; Parent *et al.*, 2006). A number of models for spatial datacubes have been proposed in recent years. However, none of these models explicitly includes integrity constraints. Consequently, integrity constraints of spatial datacubes are treated in non-systematic and pragmatic ways, which makes the verification process of logical consistency in spatial datacubes cumbersome and inefficient.

This research provides a framework for the identification of integrity constraints of spatial datacubes. To achieve this, we first propose a formal model for spatial datacubes. This model formally describes the different components of spatial datacubes, i.e., a level, a dimension, a measure, a hyper-cell, and a datacube. Based on this model, we identify and categorize different types of integrity constraints in spatial datacubes. These integrity constraints include both traditional integrity constraints (in the context of spatial datacubes) as well as cube-specific integrity constraints.

Traditional integrity constraints should be defined to resolve possible inconsistencies that arise when integrating heterogeneous spatial data into a single spatial datacube. For example, consider the level 'city' of a spatial dimension. An example of traditional integrity constraints for this level is "two cities must be either disjoint or adjacent". This traditional integrity constraint prevents inconsistent situations when, for example, the geometrical representations of two cities overlap. On the other hand, cube-specific integrity constraints are specific to spatial datacubes and include a number of sub-categories such as summarizability integrity constraints (to make data summarization meaningful), hyper-cellability integrity constraints (to avoid meaningless hyper-cells), and fact integrity constraints (to address unsound facts). For instance, since the result of aggregating the measure 'postal code' using a given aggregation function (e.g., SUM) is neither semantically meaningful nor arithmetically possible, a summarizability integrity constraint states that one must not aggregate the measure 'postal code' using an aggregation function.

Logical consistency is one of the important elements of data quality and integrity constraints play a central role in improving this element. The proposed categorization of integrity constraints provides a framework for spatial datacube designers (analyst) to systematically identify the integrity constraints early in the design stage of spatial datacube applications. Using this framework, it is not required that spatial datacube designers put significant time and effort into identifying the integrity constraints of every spatial datacube they design based on ad-hoc experiences and pragmatic ways. In addition, this classification of integrity constraints has been used as a foundation to develop a language for expressing integrity constraints of spatial datacubes.

# References

Bédard, Y., S. Rivest and M.J. Proulx. (2007), "Spatial on-line analytical processing (SOLAP): Concepts, architectures and solutions from a geomatics engineering perspective". *In:* R. Wrembel and C. Koncilia (eds.). *Data warehouses and OLAP: Concepts, architectures and solutions*, IRM Press (Idea Group), London, UK, pp. 298-319.

Godfrey, P., J. Grant, J. Gryz and J. Minker. (1998), "Integrity constraints: Semantics and applications". *In:* J., Chomicki and G. Saake (eds.). *Logics for databases and information systems*, Kluwer Academic Publishers, Boston, USA, pp. 265-306.

Parent, C., S. Spaccapietra and E. Zimányi. (2006), *Conceptual modeling for traditional and spatio-temporal applications: The MADS approach*, Springer-Verlag, Berlin, Germany, 466p.

**SECTION II – APPLICATIONS**

# 7

## 3D Building Generalization under Ground Plan Preserving Quality Constraints

*Martin Kada, Michael Peter & Yevgeniya Filippovska*

## Abstract

*In cartographic presentations, spatial objects undergo a generalization process to adjust their graphical description to the respective scale and projection. One aspect is the geometric simplification that, when performed autonomously on single objects, can lead to spatial conflicts with the entity's neighbourhood. If a displacement is not desired, or perhaps even prohibited, it is inevitable that the generalized objects remain inside their original outlines as best as possible. In this paper, we present a generalization algorithm for 3D building models, an extension that adjusts the generalized shapes to better fit their original 2D ground plan, and quality metrics that evaluate the generalization results with regard to the aforementioned requirements.*

## 7.1 Introduction

The automatic generalization of 3D building models has been a research topic for almost a decade, during which time several approaches have been proposed that simplify single objects under constraints that are specific to buildings. Among researchers, it has been agreed upon that the co-planar, parallel and rectangular alignment of facade walls is important and must be preserved by the generalization process. To accomplish an adequate simplification under these stringent conditions, generalization approaches have adapted surface simplification operators (Coors, 2001; Kada, 2002; Rau *et al.*, 2006), use morphology (Forberg, 2004), feature segmentation (Thiemann and Sester, 2004), silhouette generalization (Anders, 2005), template matching (Thiemann and Sester, 2006) and crystallographic analysis (Poupeau and Ruas, 2007). In Section 7.2, we briefly present another automatic generalization approach for 3D building models, which re-creates an object as a cell decomposition using a minimum set of planes that approximate the facade and wall polygons of the initial model. Due to the approximation process, the generalized models will converge towards their original points, which, in many cases results in ground plans that are true to their original ground areas.

A geometric simplification of 3D building models leads inevitably to differences between the original and generalized 2D contours, which will often result in conflicts between neighbouring entities. The simplification, for example, might result in buildings that overlap each other, or form gaps between formerly adjoining entities. This might lead to a misinterpretation of the spatial situation depicted in the map. In order to minimize these unwanted inconsistencies, a fitting of the generalized 3D shape to the original 2D ground plan becomes necessary. As map-like 3D presentations may focus on different aspects regarding the trueness of location, contour, area, etc., the fitting is thought of as a post-processing stage of the simplification process that can be exchanged by other algorithms that prioritize other properties. In Section 7.3, we show such an approach with a strong focus on the original contour lines.

As always, if there are two or more different ways to achieve the same goal, there is also a need for an impartial evaluation about the quality of the results. Bard and Ruas (2004) and Frank and Ester (2006) describe quality evaluation concepts for the cartographic generalization of whole maps. Podolskaya *et al.* (2007) propose an aggregated metric for polygon generalization and show its value both for 2D buildings and land cover polygons. In order to quantify the improvement that the adjustment yields, Section 7.4 discusses criteria that are specific for 2D building simplification with regard to the original contours. In Section 7.5 we show results from the generalization and quality evaluation for a number of buildings.

## 7.2  3D Building Generalization with Approximating Planes

Our generalization approach is best described by using the analogy of sculpting, where the artist creates a rough shape of the sculpture by knocking off unwanted portions from a large, solid block. However, the sculptor is restricted to a tool that cuts entirely through space, thus splitting everything in its way into two parts. Then, she peels away the pieces that shall not contribute to the result and glues the remaining parts together to form the final shape. Now, in order to create a simplified replica of a given 3D building model, the number of performed cuts must be as few as possible, but still enough to reproduce the characteristic shape of the original building. The spatial partitioning of the block into a collection of adjoining nonintersecting solids can be regarded as a cell decomposition, which is a representation in solid modeling (Foley *et al.*, 1996).

Figure 7.1 shows the five steps of the generalization process on a rather complex, horseshoe-shaped 3D model of a palace.

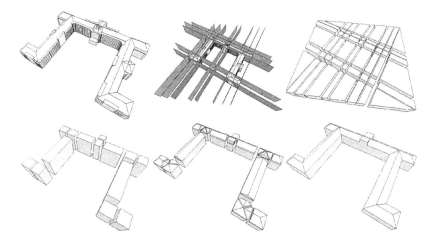

**Figure 7.1.** The detailed building model of the New Palace of Stuttgart is simplified in five steps towards its elementary shape.

In the first three steps, a 2D decomposition is generated that approximates the ground plan by a disjoint set of cells. This is achieved by deriving vertical decomposition planes from the facade walls (step 1), subdividing a large, solid block along these planes (step 2) and keeping only cells that have a large overlap with the original ground plan (step 3). Then, the three steps are repeated to generate simplified roofs for each cell from decomposition planes derived from the original roof faces (step 4). A union of the cells concludes the generalization (step 5). The major contribution is how these decomposition planes are generated.

### 7.2.1 Approximating planes

For a building model $M = \{F_1, F_2, ..., F_n\}$ of $n$ faces $F_1$ to $F_n$, we seek a partition $P = \{P_1, P_2, ..., P_m\}$ where $P_i = (B, F_P, F_{NP})$ and $m \ll n$. Each element $P_i$ of $P$ consists of a buffer $B = (A, B, C, D_{min}, D_{max})$, which is delimited by the two parallel planes $Ax + By + Cz + D_{min} = 0$ and $Ax + By + Cz + D_{max} = 0$, and two sets of faces $F_P$ and $F_{NP}$, which contain the faces that lie completely inside that buffer and are parallel and nonparallel respectively to the aforementioned delimiting planes. In terms of the generalization algorithm, we group parallel faces together that can be approximated by the plane $Ax + By + Cz + D_{avg} = 0$, for which $D_{avg}$ is the averaged point distance of all parallel faces.

Metaphorically speaking, we seek a buffer of parallel faces and compress the space into a single approximating plane, rendering all faces that are completely inside that region flat with regard to this plane. This is also the reason the nonparallel faces must not contribute to any other approximating plane.

When creating the cell decomposition, the number of cells increases with every subdivision. So, the rate of simplification is inversely proportional to the number of approximating planes. The fewer planes used, the less detailed the result will be. We exercise control over how many planes are generated with the help of two thresholds: first, the expansion of the buffer perpendicular to the delimiting planes $(D_{max} - D_{min} < \varepsilon_{distance})$ determines the minimal dimension of the generated cells, and therefore also for the faces of the final solid; second, to not generate planes with too many different orientations, the faces in $F_p$ are said to be parallel if the angle between their normal direction $\vec{n}$ and the vector *(A, B, C)* of the buffer planes are below the angle threshold $\varepsilon_{angle}$.

Once a set of approximating planes is generated, small inaccuracies in the orientation are adjusted so that as many planes as possible are parallel or rectangular to one another.

### 7.2.2 Generalization by cell decomposition

The generalization algorithm starts by computing strictly vertical approximating planes from the building model's facade faces. They are then used to generate the 2D cell decomposition of the building by subdividing a block with approximately two times the size of the building's bounding box. Only cells that feature a high overlap with the original ground plan will be processed further, all other cells are discarded. The large initial block ensures that the outer cells are very large, have little overlap with the ground plan, and are therefore securely discarded.

At this point, the generalization algorithm is extended to the third dimension, and approximating planes are generated from the remaining roof faces with arbitrary orientation. This is done globally for the whole object. To avoid an over-fragmentation, however, the decomposition itself is only done per cell. A plane will only subdivide a cell if at least one face *f* in the set $F_P$ from the buffer *B* (from which the plane was derived) at least partially lies inside that cell.

After the building cells have been identified by computing the 3D overlaps with the original building model, they are 'glued' together to form the generalized building model. While cell decomposition is not as versatile as constructive solid geometry, it is sufficient for creating all possible building shapes.

By using averaged plane equations, the resulting model will converge towards all original points and its ground plan will be true to its original area.

## 7.3  Ground plan fitting

In order to fit the 3D shape into the 2D ground plan, we assume the topology of the generalized 3D model to be fixed, but allow the facade walls to move along their normal direction. Initially, only the walls adjacent to the ground plan are altered but, through a network of distance ratios and fixed height levels, the remaining structure is adjusted to best keep the generalized 3D shape.

### 7.3.1  Analysis

First, the original model is analyzed and the approximating planes derived from the generalization process are further classified according to their adjacency to the

original ground plan. Only planes that are connected to the ground plan are considered moveable. For each of these moveable planes, a subset of nearly coplanar
faces is selected from the set of parallel faces that were accumulated in the buffer
during the generation of the approximating planes. Then the one with the best ratio
between the area sum of the faces and their distance to the plane is chosen.

In order to adjust the complete 3D building structure, the remaining non-vertical
approximating planes are set in relation to the aforementioned planes using distance
ratios that only consider the x- and y-coordinates of the intersection points (see Figure 7.2). To avoid topological errors, the slope of the roof planes has to change during the fitting, which is implemented by strictly maintaining the height levels of
every point.

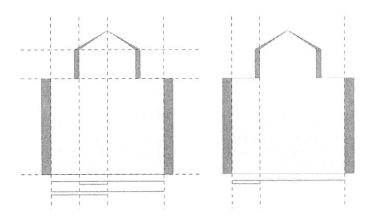

**Figure 7.2.** Distance ratios for roof (left) and wall (right) planes
that link unconnected planes to the ones adjacent to the ground
plan.

### 7.3.2  Least squares adjustment

The fitted model is obtained using least squares adjustment. In order to maintain
characteristics like rectangularity and parallelism, the approximating planes adjacent
to the ground plan are merely shifted, thus trying to minimize the distances to the
associated set of coplanar faces. As these facade faces are always vertical, the mini-
mization problem is reduced to two dimensions and we use the respective contour
lines instead of the complete faces. These lines must not necessarily be parallel to
the moving plane, as small deviations are allowed in the analysis. Because the distance between a plane and a nonparallel line is always zero, we subsample the lines
into a set of points and minimize the distance between the plane and these points.
The whole adjustment process not only fits the model to the ground plan contour, but
may also restore symmetries in the building structure that were lost in the generalization process due to the averaging (see, for example, the side wings in Figure 7.3).

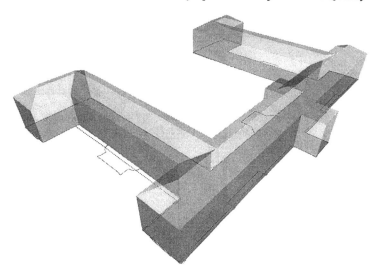

**Figure 7.3.** 3D building model fitted to the original contour (continuous line). The dashed line shows the generalized contour before the fitting process.

## 7.4 Quality Evaluation

Both in 2D maps or map-like 3D presentations, the entities need to be arranged so that no spatial conflicts arise (e.g. entities do not overlap). Although the dimensions of the various objects of a 3D city model are really 3D, their arrangement is dominantly 2D. Therefore, we limit our quality evaluation at this point to the buildings' ground plan and compare different generalization results to the original models. The following subsections discuss three quality criteria that are specific to the ground plan fitting problem.

### 7.4.1 Hausdorff distance

The first quality criterion considers ground plan contours as sets of points, which can then be compared pair wise by the maximum of all shortest point distances. This metric was first introduced by Hausdorff (1914) in his work on set theory. It helps to identify the largest geometric divergence between the original and generalized ground plan (see Figure 7.4). Although no general conclusion is possible for small Hausdorff distances, a large value signals that thin, long building parts were removed by the generalization. This happens when the distance between opposite sides of the contour is below the generalization threshold. If that thin element, for example, connects two building parts, it is essential for the building and must be preserved by all means. Otherwise the result features two objects instead of one. Under cartographic aspects, an enhancement of the respective part would be a sound solution. As our generalization approach has no means of enhancement, we can only act on large Hausdorff distances by reapplying the generalization algorithm using smaller distance thresholds.

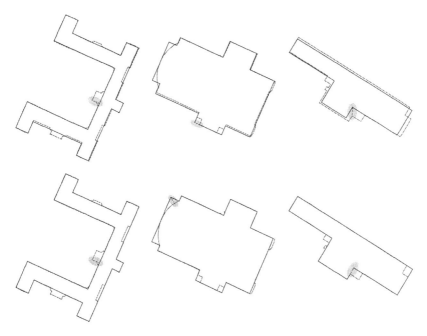

**Figure 7.4.** Hausdorff distances between original and generalized ground plan polygons before (top) and after (bottom) ground plan fitting.

### 7.4.2  Contour

The largest point deviation guarantees that all points of the generalized ground plan are closer to the original ground plan than this distance. Another criterion to look at could be the mean distance of all contour points. But it is not only interesting how much the contour changed or by how much, but also what portion of the original contour line remained the same after generalization. This can be accomplished by intersecting the contours of both ground plan polygons (see Figure 7.5).

However, building generalization algorithms tend to align wall segments in a co-planar, parallel or rectangular arrangement. So, small inaccuracies and variations will result in no overlapping contour lines and makes this quality criterion meaningless. In order to have a useful criterion, we therefore apply a small buffer around the original ground plan *(O)* and intersect it with the generalized contour lines *(G)*. To avoid that the result is slightly better because of the use of a buffer, a modified formula is used that cancels out its negative effect:

$$CT = (Length(G(Buffer(G \backslash Buffer(O))))) / (Length(O)) \quad (7.1)$$

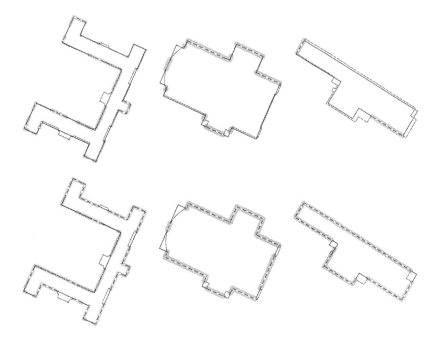

**Figure 7.5.** Contour trueness of original and generalized ground plan polygons before (top) and after (bottom) ground plan fitting.

### 7.4.3  Area differences

It is often required that a generalization should produce results that are true to area. To quantify such a change, a simple solution could be to subtract the area of the original from the generalized ground plan polygon. However, this difference does not consider the location. For example, it does not distinguish between two identical generalization results that differ only by a displacement.

So, we think it is better to also examine the space that the two ground plan polygons occupy using set theory. The intersection of the original $(O)$ and generalized $(G)$ polygons $(O \cap G)$ denotes the common region, whereas the two differences $(O \setminus G)$ and $(G \setminus O)$ are the missing and supplemental space, which together form the symmetric difference $((O \setminus G) \cup (G \setminus O))$ (see Figure 7.6). As above, the area of the symmetric difference is again set in relation to the original area. After adapting the above mentioned criteria, we obtain the following formulas for symmetric difference $(SD)$ and area difference $(AD)$:

$$SD = \frac{Area\big((O \setminus G) \cup (G \setminus O)\big)}{Area(O)} \tag{7.2}$$

$$AD = \frac{Area(G \backslash O) - Area(O \backslash G)}{Area(O)} \tag{7.3}$$

For both criteria *AD* and *SD* hold, that the smaller the value, the better the generalization. It should be noted that, according to these criteria, the best generalization is the unchanged initial ground plan polygon, as it should be, if its shortest line segment is longer than the generalization threshold. Otherwise, the simplification result does not fulfill the basic generalization requirement, and is therefore an invalid generalization of the ground plan.

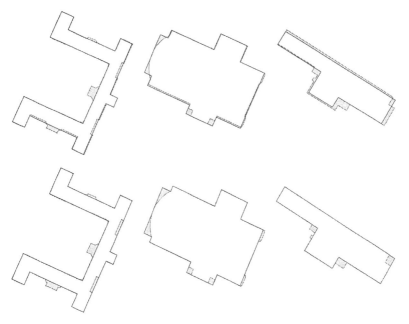

**Figure 7.6.** Symmetric difference between original and generalized ground plan polygons before (top) and after (bottom) ground plan fitting.

## 7.5  Results

The aforementioned algorithms were implemented and tested on a variety of 3D building models. For the illustration of the quality criteria, three examples have been selected that have a rather complex ground plan and which clearly depict the benefits of the ground plan adjustment (the New Palace, Opera House, and Hindenburgbau in Stuttgart; see Figure 7.4-7.6). The quality of the ground plans has been evaluated both for the generalized (G) and generalized and adjusted (G & A) versions and the results are shown in Table 7. With the exception of the contour, a lower absolute value means better quality.

When comparing the Hausdorff distance to the generalization threshold, the New Palace shows a slightly higher value. This suggests that the central entrance porch could have been included in the generalized model without introducing an element that is too small for the targeted map scale. It is also apparent that the adjustment can have the effect of increasing the Hausdorff distance, but, at least in our view, at the same time have a positive impact on the generalization quality. As already mentioned, the Hausdorff distance gives a good hint to where the generalization algorithm is too stringently applying its distance threshold.

In all cases, the contours of the building models were considerably improved by the adjustment process. This proves that the algorithm works as it is supposed to. It is also pleasing that the symmetric and area difference show better values after adjustment. However, trueness to contour and trueness to area are conflicting quality requirements and not all three criteria can be maximized at the same time.

**Table 7.1.** Comparison of the ground plan quality values for the generalized (G) and generalized and adjusted (G&A) 3D building models.

|  | New Palace | | Opera House | | Hindenburgbau | |
|---|---|---|---|---|---|---|
|  | G | G & A | G | G & A | G | G & A |
| Generalization Threshold | 10m | | 5m | | 10m | |
| Hausdorff distance | 10.65m | 10.65m | 4.32m | 5.59m | 8.52m | 7.36m |
| Contour | 28.83% | 54.75% | 34.32% | 73.43% | 14.10% | 85.35% |
| Symmetric difference | 9.64% | 7.64% | 4.48% | 3.47% | 15.29% | 4.79% |
| Area difference | -3.32% | -2.58% | 3.10% | 2.09% | -12.43% | -1.41% |

The evaluation results for a large residential area of 196 buildings can be seen in Figure 7.7-7.9. In comparison, the ground plans of these buildings are not as complex as the ones of the three landmarks. The data was generalized with a distance threshold of 2.5m. Figure 7.7 shows the original ground plans overlaid with the generalized and fitted ground plans. It is noticeable that some small buildings were completely discarded by the generalization process due to their small size. These buildings were not regarded in the analysis.

While the fitting process only slightly changes the geometry, the contours appear to be more exact afterwards. Especially the gaps between adjoining buildings fully disappeared. In this data set of simple building models, the area difference and contour trueness show the most change, so we concentrate only on these criteria.

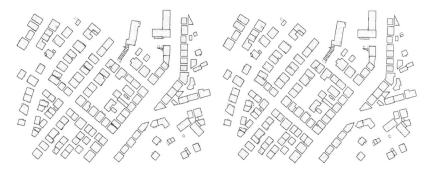

**Figure 7.7.** Original buildings overlaid by simplified ground plans (left) and overlaid by simplified and fitted ground plans (right).

When looking at the grey-scaled maps of the quality evaluation, where lighter color means better quality, it becomes apparent that the area difference (Figure 7.8) generally becomes worse by the fitting process, whereas the trueness to the contour (Figure 7.9) improves. Because the generalization algorithm generates new outlines as an average of faces, the area difference, especially, worsens for outlines with a z-shaped side: pulling an averaged line segment towards one or the other original faces shrinks or enlarges the area.

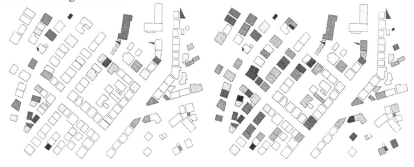

**Figure 7.8.** Area difference of generalized (left) and fitted ground plans (right).

In many cases, the fitting process improved the contour trueness. However, as the contour of the generalized outline becomes shorter, it cannot reach a perfect score anymore when comparing it to the original contour length. So, even the ones that improved by the fitting still remain darker than the perfectly fitting buildings. This suggests that it might be better to set the contour trueness in relation to the generalized contour length. This has to be looked at in future work.

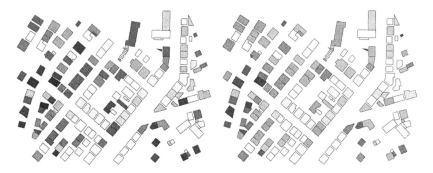

**Figure 7.9.** Contour trueness of generalized (left) and fitted ground plans (right).

## 7.6 Conclusion

The paper discusses quality evaluation under the aspect of 3D building generalization. With a strong focus on the trueness of contour, a simplification approach is briefly presented in conjunction with a post-processing fitting. In order to evaluate the improvement of the fitting or, more generally, the quality of two generalization results, we also propose four criteria that quantify a maximum point deviation, the trueness to the original contour, and the change of the ground plan area.

Applied on generalized and fitted models, the criteria clearly show that the trueness to the contour has always improved by the fitting process, while the area difference becomes worse in many cases. Both criteria are conflicting and cannot be maximized at once by the generalization. Dependent on the application, one or the other criteria might be useful in map evaluation. In future work, it will be necessary to think about how to take into account that a generalization always shortens the contour line, perhaps requiring an adjustment of the criteria.

## References

Anders, K.-H. (2005), "Level of Detail Generation of 3D Building Groups by Aggregation and Typification". *Proceedings of the XXII International Cartographic Conference (CD-ROM)*, La Coruña, Spain, 8p.

Bard, S. and A. Ruas. (2004), "Why and How Evaluating Generalised Data?" *In:* P.F. Fisher (ed.), *Developments in Spatial Data Handling. Proceedings of the 11th International Symposium on Spatial Data Handling.* Springer, NY, USA, pp. 347-342.

Coors, V. (2001), "Feature-Preserving Simplification in Web-Based 3D-GIS". *In:* A. Butz, A. Kruger, P. Oliver and M. Zhou (eds.), *Proceedings of the 1st International Symposium on Smart Graphics*, Hawthorne, NY, USA, pp. 22-28.

Foley, J., A. van Dam, S. Feiner, and J. Hughes. (1996), *Computer Graphics: Principles and Practice* (2nd edn.), Addison-Wesley, Reading, Mass., 1175p.

Forberg, A. (2004), "Generalization of 3D Building Data based on a Scale-Space Approach". *In:* O. Altan (ed.) *ISPRS Congress Istanbul 2004, Proceedings of Commission IV*, Vol 35, Part B4, pp. 194-199. http://www.isprs.org/congresses/istanbul2004/.

Frank, R. and M. Ester. (2006), "A Quantitative Similarity Measure for Maps", In: A. Riedl, W. Kainz and G.A. Elmes (eds.), *Progress in Spatial Data Handling. 12th International Symposium on Spatial Data Handling*, Springer, Berlin, pp. 435-150.

Hausdorff, F. (1914). *Grundzüge der Mengenlehre*. Verlag Veit & Co., Leipzig, Germany.

Kada, M. (2002), "Automatic Generalisation of 3D Building Models". *In:* C. Armenakis and Y.C. Lee (eds.) *International Archives of the Photogrammetry, Remote Sensing and Spatial Information Sciences, Vol. 34, Part 4.* GITC bv, Lemmer, Netherlands, pp. 243-248.

Podolskaya, E.S., K.-H. Anders, J.-H. Haunert and M. Sester. (2007), "Quality Assessment for Polygon Generalization". *In:* M. Molenaar, W. Kainz and A. Stein (eds.) *Modelling Qualities in Space and Time. Proceedings of the International Symposium on Spatial Data Quality '07, ITC, Enschede, The Netherlands.*

Poupeau, B. and A. Ruas. (2007), "A Crystallographics Approach to Simplify 3D Building". *In: Proceedings of the 23rd International Cartographic Conference (CD-ROM)*, Moscow, Russia.

Rau, J.Y., L.C. Chen, F. Tsai, K.H. Hsiao and W.C. Hsu. (2006), "Automatic Generation of Pseudo Continuous LoDs for 3D Polyhedral Building Model". *In:* A. Abdul-Rahman, S. Zlatanova and V. Coors (eds.) *Innovations in 3D Geo Information Systems*, Springer-Verlag, Berlin, pp. 333-343.

Thiemann, F. and M. Sester. (2004), "Segmentation of Buildings for 3D-Generalisation". *In: Working Paper of the ICA Workshop on Generalisation and Multiple Representation (CD-ROM)*, Leicester, UK, 7p.

Thiemann, F. and M. Sester. (2006), "3D-Symbolization using Adaptive Templates". *In: Proceedings of the GICON*, Wien, Austria, pp. 109-113.

# 8

# Assessing Geometric Uncertainty in GIS Analysis Using Artificial Neural Networks

*Farhad Hosseinali, Abbas Alimohammadi & Ali A. Alesheikh*

## Abstract

*Information on the uncertainties of GIS results is of prime importance for effective use of GIS in decision making. Development of reliable methods for representation and management of information uncertainty remains a persistent and relevant challenge to Geospatial Information Systems (GIS).*

*Artificial Neural Networks (ANNs) have increasingly been used for weighted integration of data of varying sources in GIS. In this paper, integration of spatial data by an ANN for the purpose of mapping and visualizing the uncertainty in a mineral potential mapping problem has been considered. Three simulated maps with various classes have been used. For training the network as well as finding the best network, 86 sample points have been used. Three types of uncertainty, including the uncertainty at boundaries of classes and sensitivity of results to the distribution, and number of training sample points have then been assessed.*

*Results have shown that ANN can be successfully used to map and visualize the influences of geometric uncertainties in the integrated maps. The integrated maps show the certainty levels together with the suitability values which can be very useful for more efficient use of the GIS analysis products.*

## 8.1  Introduction

In Geoinformatics, it is widely acknowledged that uncertainty arises from abstracted models of an infinitely complex geographic world. The abstracted model mainly refers to discrete representation, finite levels of detail, incomplete data collection, deficient knowledge, etc. It is often said that uncertainty is an inherent property of GIS data or geographic phenomena (Shu *et al.*, 2003). Uncertainty refers to "any deviation from the unachievable goal of completely deterministic knowledge of the relevant system" (Walker *et al.*, 2003).

In the past, there was only exclusive use of precise data or the data was assumed to be certain. In recent years, there is more allowance of 'soft data', which refers to any data with uncertainties (Aerts and Clarke, 2003). Representing data quality and uncertainty is a huge theme in the world of cartography and geographic information science. Uncertainty need not be excised as a flaw, but needs to be managed and accepted as an intrinsic part of the complex knowledge (Couclelis, 2000).

A good deal of research effort has been directed towards the study of uncertainty in geographic data (Gahegan and Ehlers, 2000; Alimohammadi *et al.*, 2004). A useful framework, recognizing the separate error components of value, space, time, consistency and completeness was proposed by Sinton (1978) and later embellished by Chrisman (1991). However, uncertainty in geographic data can be described in a variety of alternative ways, such as those provided by Bedard (1987), and Veregin (1989) which include metric sense in a space-time framework or more abstract concepts describing coverage and reliability.

Work on uncertainty to date addresses the inherent errors present within the specific types of data structure (e.g. raster or vector) or data models (e.g. field or object). The effects of combining data layers together within these various paradigms have been studied by many researchers (such as Veregin, 1989; Openshaw *et al.*, 1991; Goodchild *et al.*, 1992; Heuvelink and Burrough, 1993; Ehlers and Shi, 1997, and Alesheikh *et al.*, 1999). Scant attention has so far been given to the problem of modeling uncertainty as the data is transformed through different models of geographic space. A notable exception is the work of Lunetta *et al.* (1991).

Generic uncertainty is divided into uncertainty of entities and uncertainty of human-machine-earth relations. Uncertainty of entities is implied by the differences among entities inside the human cognition or computing machine or the earth. Uncertainty of human-machine-earth relations arises from the differences among cognitive, computational and geographic entities (Shu *et al.*, 2003).

Uncertainty can appear in any type of information. However, in GIS we often deal with spatial uncertainty which is related to spatial data. When transforming data between different conceptual models of geographic space, the uncertainty characteristics in the data may change. Unfortunately, spatial data are often analyzed and communicated with considerable and non-negligible levels of uncertainties. Uncertainty is created and propagated when we conceptualize, model, measure, analyze and represent various characteristics of the real word (Figure 8.1).

**Figure 8.1:** Sources of uncertainty in various stages of GIS operations.

Spatial sensitivity analysis, by using the different realizations of the possible errors in the input data, can be very useful for quantification of the potential prediction errors caused by the different uncertainty sources (Jager *et al.*, 2000).

## 8.2 Calibration and Weighting of Information Layers

In many GIS analysis and modeling tasks, different types of data are usually integrated. Determination of the relative importance of information is called map layer weighting (Malczewski, 1999). In general, each layer of information includes a number of sub-classes. The importance of sub-classes has to be determined before assigning weights to the layers. This procedure is called calibration and the weights assigned to the classes are called ratings (Berry, 2002). There are two main methods for weighting the information layers; data-driven and knowledge-driven (Bonham-Carter, 1994). In data-driven methods the importance of data is determined by using data itself, while in knowledge-driven methods weights are defined by an expert or a group of experts (Hosseinali and Alesheikh, 2008).

## 8.3 Artificial Neural Networks as a Weighting Method

Artificial neural networks have been used in many branches of science due to their versatile characteristics (Graupe, 2007). An artificial neural network operates by creating connections between many different processing elements, each analogous to a single neuron in a biological brain. Each neuron takes many input signals, then, based on an internal weighting system, it produces a single output signal which is typically sent as an input to the other neurons (Porwal *et al.*, 2003).

Ability of learning is one of the most important characteristics of ANN. Based on the type of training, ANNs are categorized into two main classes of supervised and unsupervised networks. The network weights are modified in the training process through a number of learning algorithms such as back propagation learning (Beale and Jackson, 1990).

A feed forward multilayer network consists of three layers: input, output, and hidden layers. Each layer in a network contains adequate number of neurons depending on the specific applications. The number of neurons in the input layer is equal to the number of data sources, and the number of neurons in the output layer depends on the application and it is usually limited by the number of output classes. The number of hidden layers and neurons in each layer depend on the architecture of network and are usually determined by trial and error (Samanta *et al.*, 2006).

ANN can be used for integrating spatial data (Kanungo *et al.*, 2006). In the process of learning, the network iteratively changes its internal weights to improve the results. Therefore, higher internal weights are assigned to the more important data. Thus, the internal weights of the network can be treated as the map layer weights. If the network contains two or more layers, a multiplication of internal weights produces the expected result.

## 8.4   Data, Materials and Methodology

In a data-driven weighting method (like supervised ANN), the simple existence of the data which must be weighted is not sufficient for performing the weighting procedure. In other words, there is a need for training samples to be used as the known outputs. Using these data, the method would attempt to recognize which data layer has more influence for prediction of the outputs. These relative effects are interpreted as weights.

To illustrate, we might assume that preparing a potential map such as mineral potential or potential of landslide is requested. In this instance, some factors are determined as effective factors and they must be integrated as predictors to produce the requested potential map. In a data-driven method, existence of some evidence or records from the regarded phenomena is required. Neural networks cannot operate with few sample points and, unfortunately, an adequate number of training samples was not accessible, therefore simulated samples were used.

Three simulated polygon maps were used as factor maps. These are assumed as predictors for a hypothetical phenomenon. Polygon maps were composed of several classes. They were represented by raster maps of 500 rows and 300 columns of 10×10 meter cells. The samples of hypothetical phenomena were a point map representing presence or absence of a phenomenon of interest. It is obvious that random propagation of sample points cannot lead to any interpretable result. Therefore, a regional pattern was considered and the points were located in relation to the assumed pattern. Finally, to add noise to the data, a few of the points were moved. The input layers are illustrated in Figure 8.2. It is expected that some polygons have major effects on explaining the existence of the hypothetical phenomenon. Based on this, the ratings of each class in the maps were determined manually with regard to the distribution of sample points (Table 8.1).

The maps should be weighted and integrated using a multi-layer perceptron artificial neural network. For this task, the most appropriate architecture for the network should first be defined by a trial and error method. So, the sample points were divided into two categories of training and check points, 56 and 30 points respectively. Back propagation was used as the learning method. In the training stage, the values of three factor maps at the position of each training point were entered into the network, and the true result for the suitability of each point was used to train the network. To achieve the best architecture for the network, the number of hidden layers and their neurons varied from 1 to 3 and 1 to 9 respectively. The criteria for choosing the best architecture were RMSE of check points, and evaluation of the network was based on check points rather than the training points. To avoid special cases, distribution of training and check points as well as values of primary internal weights of networks varied 10 times for each architecture of the network. So, for each architecture, a mean of 100 random cases was calculated. The best architecture was the one with the minimum RMSE. By doing so, the best network was composed of two hidden layers, with 8 and 7 neurons respectively. The integration of three factor maps by the selected ANN is shown in Figure 8.3. The weights assigned to each factor map by the network can be extracted from the network by multiplying the corresponding internal weights (Table 8.1).

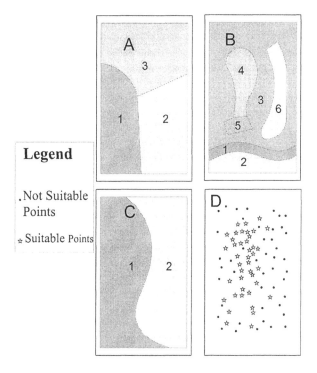

**Figure 8.2:** The input maps (A, B, C) with their classes and pattern observation points (D).

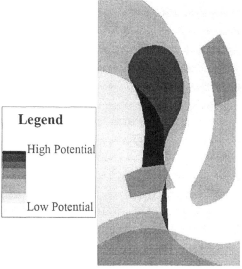

**Figure 8.3:** Map of pattern potential resulting from the supervised ANN.

**Table 8.1.** Weights and ratings of factor maps, where the ratings were assigned based on the knowledge (Knowledge driven) and the weights were extracted from the ANN (Data driven)

| Layers | A | | | B | | | | | | C | |
|---|---|---|---|---|---|---|---|---|---|---|---|
| **Weights** | 0.3 | | | 14.785 | | | | | | 14.025 | |
| **Classes** | 1 | 2 | 3 | 1 | 2 | 3 | 4 | 5 | 6 | 1 | 2 |
| **Ratings** | 0.1 | 0.5 | 0.4 | 0.05 | 0.15 | 0.01 | 0.49 | 0.16 | 0.14 | 0.9 | 0.1 |

## 8.5  Uncertainty Mapping

Three methods for examining uncertainty were used in this study. In the first test, uncertainty at the boundaries of the classes was taken into consideration. In the second and third tests, sensitivity of result to the distribution and number of sample points were assessed.

The main assumption in the modeling of geometric errors was concentrated in uncertainties of class boundaries (cf. Figure 8.4A). So, the values assigned to the class boundaries would have higher rates of uncertainties. Two different functions, including the linear (Figure 8.4B) and a Gaussian normal (Figure 8.4C), were used for definition of the variation of values between the adjacent classes. Applications of these functions were based on the definition of buffers as indicators of horizontal error around the boundaries. By using a similar network with the preferred architecture, as described in the previous section, and by considering a buffer width of 93 meters around boundaries of classes, the new values for factor maps were entered to the network and the resulting potential map was produced. Figure 8.5 shows how uncertainties at the boundaries were modeled.

**Figure 8.4:** Three models of value change between two adjacent classes. A: sharp change, B: change by a linear function, C: change by a Gaussian normal function.

In the second test, the number of training samples was kept constant (56), whereas their location was changed 100 times. By using the new data as the training samples in the network, 100 values were obtained for each cell. The standard deviation of the values assigned to each cell was calculated and used as an indicator of uncertainty of location for the sample points. The third test was mostly like the second one, but instead of changing the location of sample points, the number of sample points varied. The number of randomly selected sample points was 83 at first and they were removed randomly one by one at each stage until reduced to just 3 points. Hence, for each cell 80 values were calculated and the standard deviation of these values was used as indicator of the sensitivity of the analysis to the number of sample points.

## 8.6  Results and Discussion

The potential maps created with and without considering the geometric uncertainties are shown in Figure 8.5.

It seems that the use of linear function in this case study has produced the smoothest variation at the boundaries. So, results of the linear function were used for the rest of the analysis. In this case, we want to determine suitable (high potential) areas while taking the uncertainties into account. Hence, the potential map including the uncertainty and boundaries (Figure 8.5C) were overlapped by the maps showing the sensitivity of the analysis to the distribution and number of sample points (Figure 8.6).

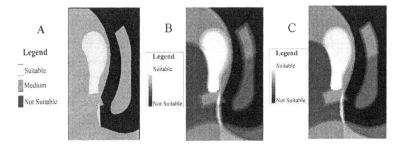

**Figure 8.5:** The potential maps without considering uncertainty (A), including the geometric uncertainty with linear (B) and Gaussian normal (C) functions.

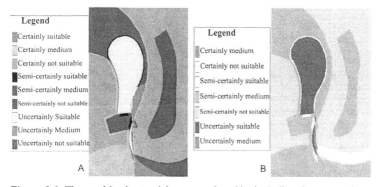

**Figure 8.6:** The combined potential maps produced by including the geometric uncertainty as well as the sensitivity to the (A) distribution of observation points and (B) number of observation points.

The produced maps could have a maximum of nine classes because each of the overlapped maps had three classes. The potential map had three classes of high, medium and low potential, and the sensitivity map had three classes of high certainty, relatively high certainty and low certainty (high uncertainty). In the case of uncertainty resulting from location of points (Figure 8.6A) all of the nine possible classes are present, while in the case of the number of observation points (Figure 8.6B) only seven classes exist. Figure 8.6 shows that the regions which are classified as suitable (high potential) areas are located in low certainty areas. Nevertheless, there are areas which are classified as not favorable or not suitable with high certainty. The reason for this situation can be related to the distribution of input points (Figure 8.2D). It is obvious that the points in the area that are classified as suitable are denser than the other areas. Therefore, when the location of points is changed randomly, lack of points in the area results in its classification as a not suitable region. On the other hand, when most of the points in this area are used, the area is classified as a suitable region. This variation results in the highest standard deviation and the highest uncertainty.

Altogether, it can be seen which regions are completely unsuitable. Nonetheless, when dealing with choosing the best area, one is never perfectly sure. One may choose the areas that are suitable, with low or medium certainty.

## 8.7  Summary and Conclusions

In this research a multi-layer perceptron neural network has been used for weighting and integrating three factor maps for the purpose of pattern prediction. 86 point samples have been used for training and choosing the best network. These points have been categorized in two classes of presence and absence of a pattern. The output or potential maps have been classified in three categories of suitable, relatively suitable and not suitable. The best network has been chosen by a trial and error approach, and this network has been then used for integration of the maps.

Two types of uncertainty have been used and overlaid on the output map. The first is the geometric uncertainty in boundaries of classes of the factor maps. Sensitivity of the output map to the changes in number and distribution of the sample points has been the second subject of consideration.

In the output maps, the suitable areas show a high rate of uncertainty, whereas unsuitable areas are identified with a minimum level of uncertainty.

It can be concluded that the distribution and number of sample points can have significant influences on the results of the ANN. Also, ANN shows high capabilities for handling the geometric uncertainties in the factor maps. A useful measure of uncertainty can be produced by combination of the geometric uncertainty and sensitivity data. Further research is recommended to include other aspects of uncertainty in GIS-based modeling.

## References

Aerts, J.C.J.H. and K.C. Clarke. (2003), "Testing popular visualization techniques for representing model uncertainty". *Cartography and Geographic Information Science,* Vol. 30(3): 249-261.

Alesheikh, A.A., J.A.R. Blais, M.A. Chapman and H. Karimi. (1999), "Rigorous Geospatial Data Uncertainty Models for GIS". *In:* K. Lowell and A. Jaton (eds.). *Spatial Accuracy Assessment: Land Information Uncertainty in Natural Resources,* Ann Arbor Press, Michigan, USA, pp. 195-202.

Alimohammadi, A., H.R. Rabiei and R. Zeaiean Firouzabadi. (2004), "A new approach for modeling uncertainty in remote sensing change detection process". *In:* S.A. Brandt (ed.). *Geoinformatics 2004: Proceedings of the 12th International Conference on Geoinformatics – Geospatial Information Research: Bridging the Pacific and Atlantic,* Gävle University Press, Sweden, pp. 503-508.

Beale, R. and T. Jackson. (1990), *Neural Computing: An Introduction,* Institute of Physics Publishing, Bristol, U.K., 240p.

Bédard, Y. (1987), "Uncertainties in land information databases". *In:* N.R. Chrisman (ed.) *Auto-Carto 8. Proceedings of the 8<sup>th</sup> International Symposium on Computer-Assisted Cartography,* Baltimore, Maryland, p. 175–184.

Berry, J.K. (2003), *Map analysis: Procedures and Applications in GIS Modeling,* Basis Press, 324p.

Bonham-Carter, G. (1994), *Geographic Information Systems for Geoscientists: Modelling with GIS.* Pergamon Press, Oxford, 398p.

Chrisman, N.R. (1991), "The error component in spatial data". *In:* D.J. Maguire, M.F. Goodchild and D.W. Rhind (eds.). *Geographical Information Systems: principles and applications,* Vol. 1, Longman Scientific & Technical, Essex, pp. 165-174.

Couclelis, H. (2000), "A Plenum Ontology of Spatial Change". *In:* A.R. Caschetta (ed.). *Proceedings of First International Conference on Geographic Information Science.* Savannah, Georgia, USA, pp. 13-14.

Ehlers, M. and W. Shi. (1997), "Error modelling for integrated GIS". *Cartographica,* Vol. 33(1): 11–21.

Gahegan, M. and M. Ehlers. (2000), "A framework for the modelling of uncertainty between remote sensing and geographic information systems". *ISPRS Journal of Photogrammetry & Remote Sensing* Vol. 55: 176–188

Goodchild, M.F., S. Guoqing and Y. Shiren. (1992), "Development and test of an error model for categorical data". *International Journal of Geographical Information Systems,* Vol. 6(2):87-103.

Graupe, D. (2007), *Principles of Artificial Neural Networks, Advanced Series in Circuits and Systems. 2nd Edition,* World Scientific Publishing Company, 250p.

Heuvelink, G.B.M. and P.A. Burrough. (1993), "Error propagation in cartographic modelling using Boolean logic and continuous classification". *International Journal of Geographical Information Systems*, Vol. 7(3): 231–246.

Hosseinali, F. and A.A. Alesheikh. (2008), "Weighting Spatial Information in GIS for Copper Mining Exploration". *American Journal of Applied Sciences*, Vol. 5(9): 1187-1198.

Jager, H.I., T.L. Ashwood, B.L. Jackson and A.W. King. (2000), "Spatial uncertainty analysis of ecological models". *Proceedings of the 4th International Conference on Integrating GIS and Environmental Modeling (GIS/EM4): Problems, Prospects, and Research Needs*, Banff, Canada, pp. 1-17.

Kanungo, D., M. Arora, S. Sarkar and R. Gupta. (2006), "A comparative study of conventional, ANN black box, fuzzy and combined neural and fuzzy weighting procedures for landslide susceptibility zonation in Darjeeling Himalayas". *Engineering Geology*, Vol. 85(3-4): 347-366.

Lunetta, R.S., R.G. Congalton, L.K. Fenstermaker, J.R. Jensen, K.C. McGwire and L.R. Tinney. (1991), "Remote sensing and geographic information system data integration: error sources and research issues". *Photogrammetric Engineering & Remote Sensing.* Vol. 57(6): 677–687.

Malczewski, J. (1999), *GIS and Multicriteria Decision Analysis*, John Wiley and Sons, New York, USA, 392p.

Openshaw, S., M. Charlton and S. Carver. (1991), "Error propagation: a Monte Carlo simulation". *In:* I. Masser and M. Blakemore (eds.). *Handling Geographic Information*, Longman Scientific & Technical, England, pp. 78–101.

Porwal, A., E. Carranza and M. Hale. (2003), "Artificial Neural Networks for Mineral-Potential Mapping: A Case Study from Aravalli Province, western India". *Natural Resources Research*, Vol. 12(3): 155-171.

Samanta, B., S. Bandopadhyay and R. Ganguli. (2006), "Comparative Evaluation of Neural Network Learning Algorithms for Ore Grade Estimation". *Mathematical Geology*, Vol. 38(2): 175-197.

Shu, H., S. Spaccapietra, C. Parent and D.Q. Sedas. (2003), "Uncertainty of Geographic Information and its Support in MADS". *In:* W. Shi, M.F. Goodchild and P.F. Fisher (eds.). *Proceedings of 2<sup>nd</sup> International Symposium on Spatial Data Quality '03*, Hong Kong, China. http://www.mics.org/getDoc.php?docid=736&docnum=1 (last accessed May 4, 2009).

Sinton, D. (1978), "The inherent structure of information as a constraint to analysis: mapped thematic data as a case study". *In:* G. Dutton (ed.). *Harvard Papers on GIS*. Vol. 7:1-19 Addison-Wesley, Reading MA, USA.

Veregin, H. (1989), "Error modelling for the map overlay operation". *In:* Goodchild, M.F. and S. Gopal (eds.). *Accuracy of Spatial Databases,* Taylor and Francis, London, U.K., pp. 3–19.

Walker, W.E., P. Harremoes, J.P. Rotmans, J.P. Van der Sluijs, M.B.A. Van Asselt, P. Janssen and M.P. Krayer von Krasuss. (2003), "Defining uncertainty - a conceptual basis for uncertainty management in model-based decisions support". *Integrated Assessment*, Vol. 4(1): 5-17.

# 9

# Neural Network Based Cellular Automata Calibration Model for the Simulation of a Complex Metropolis

*Hamid Kiavarz Moghaddam & Farhad Samadzadegan*

## Abstract

*The calibration of CA (Cellular Automata) models is very difficult when there is a large set of parameters. In the proposed model, most of the parameter values for CA simulation are automatically determined by the training of an artificial neural network. In this paper, an ANN (Artificial Neural Network) based CA urban growth simulation and prediction of the city of Esfahan in Iran over the last four decades succeeds in simulating specified tested growth years at a high accuracy level. Some real data layers have been used in the ANN-CA simulation training phase, such as the urban area in the year 1990, while others were used for testing the prediction results, such as data from 2001. The next step includes running the developed ANN-CA simulation over classified raster data for forty years. An ArcGIS extension has been developed to define the ANN-CA algorithm on a real urban growth pattern. Uncertainty analysis is performed to evaluate the accuracy of the simulated results as compared to the real historical data. Evaluation shows promising results represented by the high average accuracies achieved. The average accuracy for the predicted growth images from 1975 and 2001 is over 90%. A modification of the ANN-CA model is based on the urban growth relationship for Esfahan over time, as can be seen in the historical raster data. The study shows that the model has better accuracy than traditional CA models in the simulation of nonlinear complex urban systems.*

## 9.1 Introduction

Cellular automata models consist of a simulation environment represented by gridded space (raster), in which a set of transition rules determine the attribute of each given cell taking into account the attributes of neighbouring cells. These models have been very successful in view of their operationality, simplicity and ability to embody logic – as well as mathematics-based transition rules – in both theoretical and practical examples. Even in the simplest CA, complex global patterns can emerge directly from the application of local rules, and it is precisely this property of emergent complexity that makes CA so fascinating and their use so appealing.

Cellular automata are powerful spatial dynamic modeling techniques that have been widely applied to model many complex dynamic systems, and a variety of urban CA models have been developed to simulate either artificial or realistic cities (Batty and Xie, 1994; Clarke *et al.*, 1997; White *et al.*, 1997; Wu and Webster, 1998; Li and Yeh, 2000). Cities, like most geographical phenomena, are complex nonlinear systems involving spatial interactions which cannot easily be modeled with the functionalities of current GIS software (Batty et al., 1999).

CA-based approaches are useful in the study of urban and regional spatial structure and evolution. A critical issue in CA simulation is the provision of proper parameter values or weights so that realistic results can be generated. Real cities are complex dynamic systems that require the use of many spatial variables in CA simulation. Each spatial variable makes a contribution to the simulation, and its influence is determined by its associated parameter or weight. A variable associated with a larger parameter value usually indicates that it is more important than those with small parameter values. There are usually many parameter values to be defined within a CA model, and the results of CA simulation are very sensitive to them (Wu, 2000).

Empirical data can be used to calibrate CA models to find suitable parameter values. Calibration is important to the generation of the best fit to actual urban development, and calibration procedures have been discussed in nonlinear dynamic spatial interaction models (Lombardo and Rabino, 1986).

## 9.2  CA Calibration and Simulation Using Neural Networks

In this model, the transition rules of CA simulation are represented by the neural network calibrated by the empirical data. In general, CA models are used to estimate the conversion probability between states. For example, a simple urban CA model can be expressed by the following neighbourhood-based transition rules (Batty, 1999):

if any cell $\{x \pm 1, y \pm 1\}$ is already developed,

$$\text{then } P_d \{x, y\} = \frac{\Sigma_{i,j \in \gamma} P_d \{i,j\}}{8} \text{ , and}$$

if $P_d \{x,y\} >$ some threshold value,
then cell $\{x,y\}$ is developed with some other probability $\rho\{x,y\}$

where $P_d \{x,y\}$ is the urban development probability for cell $\{x,y\}$ and cell $\{i,j\}$ is the set of all the cells which are from the Moore neighbourhood (Clarke et al., 1997) $\gamma$ including the cell $\{x,y\}$ itself. The probability of a cell being developed is decided by the number of already developed cells in the neighbourhood. Usually, there is a higher chance of a cell being developed if it is surrounded by more developed cells.

Simulation based on a single factor of developed cells in the neighbourhood cannot address complicated urban systems. More factors have been incorporated in CA models to improve simulation performance for either hypothetical or realistic applications: distance, direction, density thresholds, and transition or mutation probabilities, for example, have been included in the transition rules of various CA models (Batty and Xie, 1994, Batty et al., 1999). Various types of constraints based on site features can also be used to regulate development patterns for land-use planning (Li and Yeh, 2000; Yeh and Li, 2001). An increase in the number of variables will result in an increase of the number of parameters in CA models. There are few concerns about the exact values of parameters if CA models are used only for hypothetical studies. However, when CA models are applied to the simulation of real cities, suitable parameter values have to be determined through some calibration procedures. The ANN-CA model consists of two separate parts: using the neural network to obtain the parameter values automatically based on training data; and using the neural network to carry out CA simulation based on these parameter values (Figure 9.1; Soares-Filho *et al.*, 2002). The first part involves the calibration (training) procedure to obtain optimal weights using empirical data. Remote sensing data are used to provide empirical data about urban development. GIS analysis is needed to obtain site attributes (attractiveness) for each cell. Neural networks can be used to reveal the relationships between development probability and site attributes. The training procedure should be outside the simulation process for computation efficiency. The parameters obtained from the training will be input to the CA model. The second step is to carry out CA simulation, which is also based on the algorithm of neural networks. At each iteration, neural networks will determine the development probability which is subject to the input of site attributes and weights. A cell may have *n* site attributes (variables).

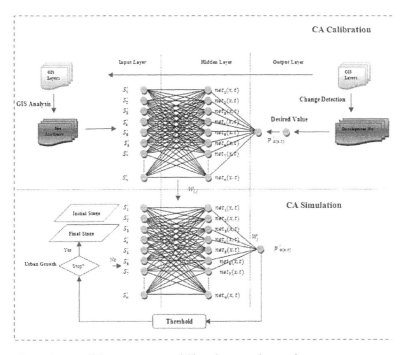

**Figure 9.1.** A cellular automaton model based on neural networks.

Urban growth is dependent on these variables, which may include various types of proximities, amount of development, and site conditions. A regression model or MCE method may not be the best way to reveal relationships because urban systems involve complex nonlinear processes. Instead, neural networks can be designed to estimate the development probability of each iteration of the CA simulation. The neural network may have three layers: one input layer, one hidden layer, and one output layer. The input layer has *n* neurons corresponding to the *n* variables. The hidden layer may also have neurons. The output layer has only one neuron which indicates development probability.

At each iteration, the site attributes of a cell will be input into the first layer and the neural network will determine its development probability at the output layer. Experiments indicate that it will be more appropriate for all original data to be converted into a range from 0 to 1 before they are input into neural networks (Gong, 1996). This is similar to data normalization in that it uses maximum and minimum values in scaling the original data set. Scaling variables treats them as equally important inputs to neural networks and makes them compatible with the sigmoidal activation function that produces a value between 0 and 1. The following linear transformation is used (Equation 9.1):

$$S_i' = \frac{S_i - minimum}{maximum - minimum}$$

(9.1)

The algorithm for the CA model is devised using a neural network. In the neural network, the signal received by neuron $j$ of the hidden layer from neuron $i$ of the first input layer for cell $x$ is calculated from Equation 9.2:

$$net_j(x,t) = \sum_i W_{i,j} S_i'(x,t)$$

(9.2)

where $x$ is a cell, $net_j(x,t)$ is the signal received for neuron $j$ of cell $x$ at time $t$, and $S_i'(x,t)$ is the site attributes for variable (neuron) $i$. The activation of the hidden layer for the signal is Equation 9.3:

$$\frac{1}{1 + e^{\left[-net_j(x,t)\right]}}$$

(9.3)

The development probability ($P_d$) for cell $x$ is then calculated from Equation 9.4:

$$P_d(x,t) = \sum_j W_j \frac{1}{1 + e^{\left[-net_j(x,t)\right]}}$$

(9.4)

A stochastic disturbance term can be added to represent unknown errors during the simulation. This can generate patterns that are closer to reality. The error term (RA) is defined in Equation 9.5 (White et al., 1997):

$$RA = 1 + (-Ln\gamma)^\alpha$$

(9.5)

where $\gamma$ is a uniform random variable within the range 0 to 1, and $\alpha$ is the parameter controlling the size of the stochastic perturbation; $a$ can be used as a dispersion factor in the simulation. The development probability is revised as Equation 9.6:

$$P_d'(x,t) = RA \sum_j W_j \frac{1}{1 + e^{\left[-net_j(x,t)\right]}} = 1 + (-Ln\gamma)^\alpha \sum_j W_j \frac{1}{1 + e^{\left[-net_j(x,t)\right]}}$$

(9.6)

Finally, a predefined threshold value is used to decide whether a cell is to be developed or not according to the probability $P_d^r(x,t)$ at each iteration. If a cell has a probability greater than the threshold value, it will be converted. The number of already developed cells in the neighbourhood is recalculated and the site attributes are updated at the end of each iteration. The number of iterations is determined by the total land consumption in a given time period. The simulation will stop when all the required number of cells has been converted into urban development.

## 9.3 Implementation and Results

This section describes the design and implementation of the ANN-CA algorithm to calibrate and simulate the urban growth of real city, namely Esfahan in Iran. The simulation process of a real city will go through many processes starting from analyzing the area of study, data processing, algorithm design and implementation, and finally evaluation of the simulation results.

### 9.3.1 Data preparation and image classification

The data that has been used for calibration and simulation included three historical satellite images covering a period of forty years. The raw images were one 60m resolution Landsat MSS image (from 1975) and four 30m resolution Landsat TM images (from 1990 and 2001), which were resampled to 60m to enable comparison. The images were geometrically rectified to the same projection of Universal Transverse Mercator (UTM) Zone 39N. Projected images were registered to spatially fit over each other using a second order polynomial transformation function and 15 well-defined control points. Registration errors were very small, represented by values far less than one pixel. After the images were geometrically rectified and registered spatially to each other, the next step was to prepare the images as inputs to the CA simulation algorithm's image classification. Six classes are defined based on a maximum likelihood classification system of the 1976 image: water, road, commercial, green area where land has been covered by forest, residential areas, and nonurban areas. Commercial and residential classes are combined after the simulation as a single 'urban' class. Ground reference data sources include classification maps derived from orthophotos, which are used for identifying the land cover classes and for training and testing data collection. The maximum likelihood classification method is used for supervised classification. A classification accuracy report is prepared for each classified image using the test data to check the quality of classification. Results indicate a high accuracy level of classification above 75% (Table 9.1).

**Table 9.1.** Classification Error Matrix

|            | Urban | Road | Green Area | Non-Urban | Water |
|------------|-------|------|------------|-----------|-------|
| Urban      | 1118  | 55   | 58         | 82        | 26    |
| Road       | 18    | 518  | 51         | 105       | 48    |
| Green Area | 0     | 0    | 630        | 51        | 52    |
| Non-Urban  | 91    | 231  | 74         | 1781      | 42    |
| Water      | 68    | 89   | 32         | 0         | 105   |

### 9.3.2 The platform of calibration and simulation

After appropriate parameter values have been obtained, they are imported into the CA model for urban simulation. The simulation is still based on neural networks. The actual simulation model is implemented in a GIS platform by the integration of the neural network, CA, and GIS. The model was programmed in ArcGIS using ArcObjects. Studies show that some distance-based variables are closely related to urban development in the region and can be used for CA simulation (Wu and Webster, 1998; Li and Yeh, 2000; Yeh and Li, 2001). Satellite images can also indicate that development sites are usually located along major transportation lines or around an existing urban area. In this study, seven spatial variables were defined to represent the site attributes of each cell for the simulation of urban development:

(1) Distance to the major (city proper) urban areas $S_1$;
(2) Distances to suburban (town) areas $S_2$;
(3) Distance to the closest road $S_3$;
(4) Distance to the closest expressway $S_4$;
(5) Distance to the closest water $S_5$;
(6) Amount of development in the neighbourhood $S_6$;
(7) Agricultural suitability (Green Area) $S_7$.

### 9.3.3 Artificial neural network based CA calibration (training)

The classification images from 1975 and 1990 (shown in Figure 9.2) were overlaid with the seven layers of site attributes. The overlay provides the training data that can reveal the relationship between site attributes and development probability. A sampling procedure was carried out by using a stratified sampling method which ensures that sampling effort can be distributed in a rational pattern so that a specific number of observations are assigned to each category to be evaluated. There were 1000 random sampling points, generated by ArcGIS, which were used to train the neural network.

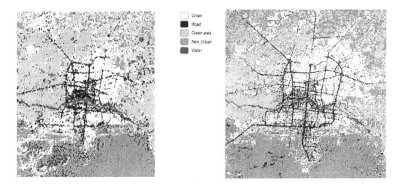

**Figure 9.2.** Ground truth of classification image for 1975 (Left) and 1990 (Right).

Table 9.2 shows examples of the training data set and the calculated development probability from the ArcGIS environment.

**Table 9.2.** Examples of parameters, desired values (actual), and calculated development probability from ANN

| $S_1$ | $S_2$ | $S_3$ | $S_4$ | $S_5$ | $S_6$ | $S_7$ | Desired value | ANN output value |
|---|---|---|---|---|---|---|---|---|
| 305 | 20 | 2 | 256 | 21 | 22 | 0.2 | 1 | 0.874 |
| 271 | 13 | 2 | 140 | 18 | 16 | 0.4 | 0 | 0.076 |
| 275 | 5 | 0 | 18 | 250 | 17 | 0.6 | 1 | 0.651 |
| 131 | 8 | 1 | 56 | 33 | 26 | 0.4 | 1 | 0.865 |
| 93 | 25 | 1 | 110 | 34 | 9 | 0.4 | 0 | 0.059 |

### 9.3.4  Artificial neural network based CA simulation

The simulation was based completely on neural networks. At each iteration, the development probability of each cell was obtained from the neural network. Cells with a development probability greater than a predefined value were converted into developed cells. Experiments indicate that a predefined value of 0.75 can produce better simulation results, but that higher values can significantly increase simulation time.

First, we simulated the urban development from 1975 to 1990 using the proposed neural-network-based model. This can be compared with the actual urban development that is detected from remote sensing. Different values of parameter α can be used to explore possible urban forms that are related to uncertainties. We implemented the algorithm with α = 1-3. The actual development is quite dispersed, however, with α = 1, results indicate that the actual land development in the region was influenced by some unexplained variables. It is also confirmed by other studies that the region is characterized by a chaotic development pattern because of severe land speculation (Yeh and Li, 2001). The second step was to predict possible future urban development from 1990 to 2001 for planning purposes (Figure 9.3).

Due to the good results achieved by ANN CA-based simulation with α = 1, we decided to simulate the growth from 1990 to the year 2001. Table 9.3 summarizes the urban growth prediction evaluation results based on ANN based CA for year 2001.

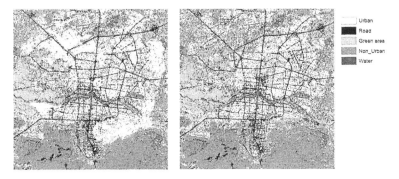

**Figure 9.3.** The ground truth (left) and ANN based CA simulated (right) images for 2001.

**Table 9.3.** ANN based CA simulation evaluation results of year 2001

| Region | 2001 Ground Truth (Urban) | 2001 Simulated (Urban) | Error | Accuracy (%) |
|--------|---------------------------|------------------------|-------|--------------|
| Esfahan | 103070 | 110960 | 7890 | 92.34 |

The accuracy achieved for the predicted year 2001 of 92.34% is encouraging for such long term prediction of an 11-year interval. Based on the ANN-based CA simulation strategy we used to predict 2001, it is likely that such prediction will help municipalities identify future growth trends and design sustainable infrastructure plans to accommodate them.

## 9.4 Conclusion

A neural-network-based CA model is better at simulating complex nonlinear systems. An ANN-CA model means that the transition rules of CA simulation are represented by the neural network. This means that users do not need to design transition rules. In traditional CA models, users may face great difficulties in choosing transition rules, so a variety of transition rules have been proposed from different CA models.

The model implemented in this paper can significantly reduce the requirements for explicit *a priori* knowledge in identifying relevant criteria, assigning scores, and determining criteria preference. Variables used in spatial decisions are very often interdependent. ANN-based CA urban growth simulation and prediction of Esfahan city over the last four decades succeeds in simulating specified tested growth years at a high accuracy level, however the interval time of the satellite images has a strong effect on predicate accuracy. The result of ANN-based CA simulation has greater accuracy (92.34%) than traditional CA models. Some ground truth images were used in the CA simulation training phase, such as the image from 1990, while the image from 2001 was used for testing the prediction results. Calibrating the ANN-based CA growth rules is important through comparing the simulated images with the ground truth to obtain feedback. An important notice is that ANN-based CA need also to be modified over time to adapt to the urban growth pattern. The evaluation method used on a regional basis has its advantage in covering the spatial distribution component of the urban growth process. These investigations show that Esfahan has a noticeable growth in urban area, and it can be anticipated that the future rate of urban growth will be more rapid than previous years.

## References

Batty, M. and Y. Xie. (1994), "From cells to cities". *Environment and Planning B: Planning and Design*, Vol. 21: 31-48.

Batty, M., Y. Xie and Z.L. Sun. (1999), "Modeling urban dynamics through GIS-based cellular automata". *Computers, Environment and Urban Systems*, Vol. 23: 205–233.

Clarke, K.C., S. Hoppen and L. Gaydos. (1997), "A self-modifying cellular automaton model of historical urbanization in the San Francisco Bay area". *Environment and Planning B: Planning and Design*, Vol. 24: 247-261.

Gong, P. (1996), "Integrated analysis of spatial data from multiple sources: using evidential reasoning and artificial neural network techniques for geological mapping". *Photogrammetric Engineering and Remote Sensing*, Vol. 62: 513-523.

Li, X. and A.G.O. Yeh. (2000), "Modeling sustainable urban development by the integration of constrained cellular automata and GIS". *International Journal of Geographical Information Science*, Vol. 14: 131-152.

Lombardo, S.T. and G.A. Rabino. (1986), "Calibration procedures and problems of stability in nonlinear dynamic spatial interaction modeling". *Environment and Planning A*, Vol. 18: 341-350.

Soares-Filho, B.S., G.C. Cerqueira and C.L. Pennachin. (2002), "DINAMICA – a stochastic cellular automata model designed to simulate the landscape dynamics in an Amazonian colonization frontier". *Ecological Modeling*, Vol. 154: 217-235.

White, R., G. Engelen and I. Uljee (1997), "The use of constrained cellular automata for high-resolution modeling of urban land-use dynamics". *Environment and Planning B: Planning and Design*, Vol. 24: 323-343.

Wu, F. (2000), "A parameterized urban cellular model combining spontaneous and self-organizing growth". *In:* P. Atkinson and D. Martin (eds.) *GIS and Geocomputation*, New York, Taylor and Francis, pp. 73-85.

Wu, F. and C.J. Webster. (1998), "Simulation of land development through the integration of cellular automata and multicriteria evaluation". *Environment and Planning B: Planning and Design*, Vol. 25: 103-126.

Yeh, A.G.O. and X. Li. (2001), "A constrained CA model for the simulation and planning of sustainable urban forms by using GIS". *Environment and Planning B: Planning and Design*, Vol. 28(5): 733-753.

# 10

## Accuracy Assessment in an Urban Expansion Model

*Amin Tayyebi, Mahmoud Reza Delavar, Bryan C. Pijanowski and Mohammad J. Yazdanpanah*

### Abstract

*Urban Expansion Models (UEM) are methods combining Geographic Information System (GIS), Artificial Neural Networks (ANNs) and Remote Sensing (RS) for urban expansion simulation. The conventional Relative Operating Characteristic (ROC) method for map comparison usually analyzes pixels at a single default scale and assigns each pixel to one category only. The purpose of this paper is to offer improved ROC methods based on different cross tabulation matrix to calibrate the UEM so that scientists can obtain more helpful results by performing multiple resolution analysis on pixels that belong to several categories simultaneously. We examine the conceptual foundation of the ROC method based on cross tabulation matrix, and then show how to extend those concepts to compare entire maps at multiple spatial resolutions. We offer a fourth improved ROC method that is designed for calibration of the UEM involving the comparison of two maps that show the same set of categories by a scientist. Results show that the methods can produce extremely different measurements, and that it is possible to interpret the differences at multiple resolutions in a manner that reveals patterns in the maps. We describe the concepts using simplified examples, and then apply the methods to characterize the change in urban land use between 1988 and 2000 in Tehran, Iran.*

### 10.1  Introduction

The conventional Relative Operating Characteristic (ROC) method of map comparison frequently produces unhelpful results for a variety of reasons. These approaches are important, nevertheless, because the cross tabulation matrix is the basis for numerous popular measurements of spatial accuracy. Improved ROC methods show the range of possibilities for constructing a cross tabulation matrix based on possible variations in the spatial arrangement of the categories within a single pixel. Scientists need to think more deeply about how to compare a single pair of pixels that have partial membership to multiple categories, because this information is necessary to understand the level of certainty we place in maps (Foody and Atkinson, 2002). Cross tabulation matrices are used to measure the spatial accuracy of raster maps (Monserud and Leemans, 1992; Congalton and Green, 1998; Wilkinson, 2005) and to quantify the association between two categorical maps. The Land Transformation model (LTM) is a land use change model that uses ANNs (Artificial Neural Networks) and GIS (Geographic Information Systems) (Pijanowski *et al.*, 2002; 2005). Scientists need a better tool to calibrate the Urban Expansion Model (UEM), because it is essential to know a model's accuracy. The categories in each contingency table are actual change and actual non-change versus simulated change and simulated non-change. Eight calibration metrics are used to estimate model goodness-of-fit: four location-based measures and four patch metrics (Pijanowski *et al.*, 2006). Cross tabulation matrices are used regularly to measure the spatial accuracy of raster maps, and more generally to quantify the association between two categorical maps for a variety of reasons (Pontius *et al.*, 2004a; 2004b; Pontius and Spencer, 2005).

If each pixel position in the study area is classified as exactly one category in the reference map and exactly one category in the simulated map (hard classified), then the pixel position is tallied in a single obvious column and row within the matrix. One reason why scientists are tempted to classify each pixel in the map as exactly one category is because the statistical analysis of such data is straightforward. However, the decision to hard classify the data introduces a fundamental problem because scientists frequently know that each pixel really contains partial membership to multiple categories. Consequently, the data processing step to hard classify the pixels corrupts the underlying information in the maps. Nevertheless, scientists regularly classify each pixel as its single dominant category, because there are not widely recognized intuitive statistical methods available to compare maps that have partial membership to multiple categories. Specifically, if the pixels do not belong to exactly one category (soft classified), then the construction of the cross tabulation matrix is not immediately obvious (Pontius, 2002; Pontius and Cheuk, 2006).

This paper presents UEM which parameterized and explores how factors such as road, built-up area, service centers, green space, elevation, aspect and slope can influence urban expansion for Tehran Metropolitan Area (TMA). The goal of this paper is to propose useful intuitive general statistical methods to compare two raster maps where the pixels can have partial membership to multiple categories. Our proposed ROC methods rely on the cross tabulation matrix, also known as the contingency table to calibrate the UEM. The structure of this paper is as follows: materials and methods provide the basic principles of ANNs as applied to the UEM, and present a complete description of proposed ROC metrics. The results and discussion present a brief summary of the study area, the results of ROC metrics for calibration of the UEM in TMA and compares four proposed ROC methods in more detail.

## 10.2  Materials and Methods

The UEM is similar to other grid-based reduced statistical models (Pontius, 2002; Walker, 2003). GIS are used to calculate a variety of spatial relationships between drivers of change and cells that could undergo change. Our UEM follows eight sequential steps: (1) rectification and registration; (2) classification; (3) integrating topographic data in a database; (4) coding of data to create spatial layers of predictor variables; (5) applying spatial or non-spatial function in ArcGIS; (6) integrating all input grids; (7) calibration of UEM; and (8) temporal prediction (Tayyebi *et al.*, in press).

When the map of a simulated urban expansion is overlaid on the reference map, a contingency table can summarize the results, as in Table 10.1. The rows of Table 10.1 show categories of the map of the model's output and the columns show the categories of the reference map. The ROC is derived from a standard two-by-two contingency table (Pontius, 2002) created when simulated and reference maps are compared. The ROC curve plots the rate of true positive to positive classifications against the rate of false positive to negative classifications as the discrimination or threshold value varies between 0 and 1. The ROC statistic is the area under the curve that connects the plotted points. This paper describes two methods for calculating the ROC. The first method involves using integral calculus' trapezoidal rule to compute the area with equation 10.1, where $x_i$ is the rate of false positives for scenario $i$, $y_i$ the rate of true positives for scenario $i$, and $n$ the number of suitability groups (Pontius and Schneider, 2001). The second approach takes advantage of a non-parametric approximation using SPSS (SPSS Inc, 2003) to estimate the area under the curve that is produced by varying the threshold. We take advantage of a second approach to estimate the area under the curve for our proposed ROC methods.

**Table 10.1.** Confusion matrix for model performance for the Kappa coefficient calculation.

| Observed | Simulated Model Run | | |
|---|---|---|---|
| change | 0 | 1 | Total |
| 0 | $P_{11}$ | $P_{12}$ | $P_{1T}$ |
| 1 | $P_{21}$ | $P_{22}$ | $P_{2T}$ |
| Total | $P_{T1}$ | $P_{T2}$ | 1 |

$$\sum_{i=1}^{n} [x_{i+1} - x_i][y_i + y_{i+1} - y_i / 2] \qquad (10.1)$$

For the first of the four proposed methods, the ROC uses a Least matrix as the contingency table. The entries in the Least matrix give the least possible association between the categories in the reference map and the categories in the simulated map. We consider possible arrangements of the categories within pixel $n$ such that category $i$ in the simulated map overlaps as little as possible with category $j$ in the reference map. If the sum of the pair of memberships is less than 1, then it is possible to arrange the categories within the pixels such that there is no overlap. If the sum of the pair of memberships is greater than 1, then category $i$ in the simulated map must have some positive overlap with category $j$ in the reference map. Equation (10.2) gives the least possible association between the simulated categories $i$ and the actual category $j$ within map $n$. We consider a different possible rearrangement for each entry in the matrix, so it is possible that the sum of all values in the matrix is less than 1 (Pontius and Connors, 2006).

$$L_{gnij} = MAX(0, X_{gni} + Y_{gnj} - 1) \qquad (10.2)$$

The second of four proposed ROC methods uses a Random matrix as the cross tabulation matrices. The entries in the Random matrix give the statistically expected association between the categories in the reference map and the categories in the simulated map, assuming those categories are distributed randomly within the maps. The membership to each category is the proportion of that category contained in the map, so the amount of overlap between a pair of categories is the mathematical product of the memberships of those categories (Pontius and Connors, 2006). Equation 10.3 gives the expected association between simulated category $i$ and actual category $j$ within map $n$, assuming random distribution of the categories within map $n$. Equation 10.3 produces the Random matrix such that the sum of all values in the matrix is equal to 1. Lewis and Brown (2001) propose an equivalent procedure.

$$R_{gnij} = X_{gni} \times Y_{gnj} \qquad (10.3)$$

In the third proposed method, ROC uses a Most matrix as the contingency table. The entries in the Most matrix give the greatest possible association between the categories in the reference map and the categories in the simulated map. We consider possible arrangements of the categories within map $n$ such that category $i$ in the simulated map overlaps as much as possible with category $j$ in the reference map. The maximum overlap is constrained by the minimum of the memberships to the two categories (Pontius and Connors, 2006). Equation 10.4 gives the greatest possible association between comparison category $i$ and reference category $j$ within map $n$. We consider a different possible rearrangement for each entry in the matrix, so it is possible that the sum of all entries in the Most matrix is greater than 1. Binaghi *et al.* (1999) propose this same minimum rule for all entries in the matrix.

$$M_{gnij} = MIN(X_{gni}, Y_{gnj}) \qquad (10.4)$$

The fourth proposed ROC method uses a Composite matrix as the cross tabulation matrices. The entries of the Composite matrix are computed according to a mathematical rule that is a composite of the concepts expressed by the minimum rule of Equation 10.4 and the multiplication rule of Equation 10.3. The Composite matrix is designed specifically for the special case where the categories in the reference map are the same as the categories in the simulated map (Pontius and Connors, 2006). For this common situation, the matrix is square, where the diagonal entries show agreement between the categories, while the off-diagonal entries show disagreement. Equation 10.4 gives the diagonal entries, i.e. when $i = j$. Equation 10.5 gives the off-diagonal entries, i.e. when $i \neq j$. Pontius and Cheuk (2006) give the full derivation of the Composite matrix. The Composite matrix has conceptual advantages over the other three matrices when the reference map and the simulated map both show the same categorical variable (Kuzera and Pontius, 2004). Specifically, when a map is compared to itself, the Composite matrix gives zeroes for all off-diagonal entries, which is not necessarily the case for the Random and Most matrices. Furthermore, the entries of the Composite rule sum to 1, which is not necessarily the case for the Least and Most matrices.

$$C_{gnij} = \left[ X_{gni} - MIN(X_{gni}, Y_{gnj}) \right] \times \frac{\left[ Y_{gnj} - MIN(X_{gni}, Y_{gnj}) \right]}{\sum_{j=1}^{J} \left[ Y_{gnj} - MIN(X_{gni}, Y_{gnj}) \right]}$$

$$(10.5)$$

## 10.3  Results and Discussion

The Tehran Metropolitan Area (TMA) is an area located in the North of Iran which exhibited accelerated rates of urban expansion over the last three decades. Tehran with a day population of 10 million and with a metropolitan area of over 2000km$^2$ is the center of the national government and commercial, financial, cultural and educational activities in Iran. Rapid land-use changes in TMA resulted from a high population growth rate and increased rural-urban migration combined with a strong tradition of centralization in the capital. The National Cartographic Centre (NCC) database was used as the main source of land-use data for TMA. Two LANDSAT ™ images of TMA with 28.5m resolution taken in 1988 and 2000 were used. The NCC database with a scale of 1:25000 was used as the source of topographic data. The locations of service centers were obtained from published county road maps.

Data on land-use were incorporated into the ArcInfo 9.2 software. Different layers in ArcMap were stored in grid format with each location containing its spatial configuration value from each driving variable grid. Effective parameters and criteria required are listed as follows (Tayyebi *et al.*, 2008a; 2008b; in press):

- **Absorbing Excursion Spaces:** Absorbing excursion spaces contain distance from service center, green space and built-up area.
- **Transportation:** Another important factor is the distance of each cell from the nearest road cell calculated and stored in a separate coverage.
- **Landscape Features:** Elevation is important in landscape that is prone to flooding. Slope and aspect are important to developers who want to minimize landscaping costs.
- **Constraint:** There are two constraints for simulation of urban expansion; (1) cells that are urban in 1988 are obviously not candidates for urban expansion in 2000; (2) cells that are protected legally from urban expansion are assigned the absolute lowest suitability value in the final suitability maps. These constraints are applied to the suitability maps created by ANN method.

The neural network toolbox of Matlab software was utilized for training and testing. We allowed the neural net to train on input and output data for 5000 cycles and saved the network file at 100 cycle intervals. It was designed to have a flexible number of inputs depending on the number of predictor variables presented to it, an equal number of hidden units as input units and an output unit. Driving variable grids stored in GIS were converted to a tabular format such that each location contained its spatial configuration value where each location was an input vector into the neural net from each driving variable grid. There is a vector of 7 by 1 measurements as input for each cell in the study area in 1988. The output layer contained binary data that represented whether a cell location changed to urban (1 = change; 0 = no change) during the study period (1988 to 2000 for TMA). The network file generated from the training exercise was used to estimate output values for each location. The output was estimated as values from 0 (not likely to change) to 1 (likely to change); the output file created from this testing exercise is called a result file; we call these suitability values. We used a maximum likelihood rule (Pijanowski *et al.*, 2002) such that cells with the greatest values were assumed to transition first. The number of cells selected to transition equalled the number of cells transitioning between the two years being modeled. A routine was written into the Java programs that performed this calculation.

In calibrating the UEM, cells that were simulated to transition to urban areas were compared with the cells that actually did transition. The study area includes 245,588 cells in which 85,956 (35.0%) have limitations to undergo transition while 159,632 (65.0%) can be subjected to transition in TMA. The ANN estimated 127,706 (80.0%) of qualified cells had changed likelihood values of 0.0, 3,193 (2.0%) had likelihood values of 1.0, and other cells 28,733 (18.0%) have a likelih-ood value between 0.0 to 1.0. Cells with values closest to 1.0 were selected as loca-tions most likely to transition. Only 3.5% of all qualified areas changed to urban in the observed databases. Results show that only 5,587 cells undergo transitions in TMA. We calculated the ROC based on the fourth method of contingency tables for each of the simulations and plotted the ROC as a function of the simulation training cycle and percentage of land use change (Figure 10.1). Figure 10.1 shows the ROC computation at 100 cycle intervals based on Least, Random, Most, and Composite matrices that compares the simulated map to the reference map for the TMA at the 30m resolution in 2000. The number of cells that were selected to train and test for simulations were 95,487 cells with a change likelihood value of 0 to 1 (urban and non-urban areas). We allowed the neural net to train on the input and output data for 5000 cycles and saved the network file at 100 cycle intervals for simulations.

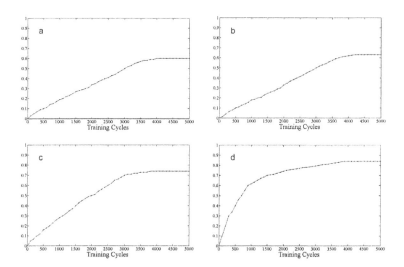

**Figure 10.1.** Accuracy assessment in UEM across training cycles: (a) ROC based on Least Matrices (b) ROC based on Random Matrices (c) ROC based on Most Matrices and (d) ROC based on Composite Matrices.

After investigation of the ROC plots based on different contingency tables, the following conclusions have been derived. ROC computation based on Least, Ran-dom, Most, and Composite matrices produce different results. The diagonal entries show persistence in the landscape while the off-diagonal entries show differences. Our four proposed ROC methods give entries that are expressed as proportions of the study area. The sum of the entries is 1 for the Random and Composite matrices, therefore those two matrices show plausible internally consistent complete descriptions of the associations between the maps. The Random matrix is the only one of the four matrices that produces entries that sum to 1 when comparing maps that show different categories. The Composite matrix requires that the categories in the reference map be identical to the categories in the simulated map. The sum of the entries is not necessarily 1 for the Most and Least matrices, so the interpretation of those two matrices must be somewhat different than for the Random and Composite matrices. The entries for the Most and Least matrices should be interpreted one entry position at a time, because each entry is computed according to a potentially different arrangement of the categories within the pixels, which is why the entries do not sum to 1. It can be useful to subtract the entries of the Least matrix from the corresponding entries in the Most matrix in order to construct the range for each entry in the matrix. The accuracy of the four proposed ROC models with maximum cycles were compared against each other and it was concluded that the rank order of the models' performance was changed respectively: (1) ROC based on Least Matrices (2) ROC based on Random Matrices (3) ROC based on Most Matrices (4) ROC based on Composite Matrices.

## 10.4  Conclusion

This paper demonstrates an adoption of the Urban Expansion Model which combines GIS, Artificial Neural Networks and Remote Sensing for TMA and explores how factors such as road, built-up area, service centers, green space, elevation, aspect and slope can influence urban expansion. Because we were interested in how well the neural networks performed against training cycles for UEM, we created different versions of the ROC metric to calibrate the UEM. This paper proposes an improved ROC approach to compare pixels that contain partial membership to multiple categories, i.e. pixels that are soft classified. It then extends the approach to compare maps at multiple resolutions. We trained the UEM for 5000 cycles, saving network files for every 100 cycle iterations. In general, it is suggested that four proposed ROC metrics are useful for calibration of UEM. It is found that more training cycles did not necessarily produce the best goodness-of-fit. Four proposed ROC accuracy of models with maximum cycles were compared against each other and it was concluded that the rank order of models performance was changed respectively: (1) ROC based on Least Matrices (2) ROC based on Random Matrices (3) ROC based on Most Matrices (4) ROC based on Composite Matrices.

The methods can be used to examine uncertainty in maps as a function of the patterns in the maps, whether or not the two maps share the same categories. Uncertainty concerning a particular categorical association is larger where the Most and Least matrices show a larger range. Results show that the ROC methods can produce extremely different measurements, and that it is possible to interpret the differences at multiple resolutions in a manner that reveals patterns in the maps. ROC computation based on Least, Random, Most, and Composite matrices produce different results. All four ROC methods give entries that are expressed as proportions of the study area. The sum of the entries is 1 for the Random and Composite matrices. The Random matrix is the only one of the four matrices that gives entries that sum to 1 when comparing maps that show different categories. The sum of the entries is not necessarily 1 for the Most and Least matrices, so the interpretation of those two matrices must be somewhat different than for the Random and Composite matrices. It can be useful to subtract the entries of the Least matrix from the corresponding entries in the Most matrix in order to construct the range for each entry in the matrix.

# References

Binaghi, E., P. Brivio, P. Ghezzi and A. Rampini. (1999), "A fuzzy set-based accuracy assessment for soft classification". *Pattern Recognition Letters*, Vol. 20: 935-948.

Congalton, R.G. and K. Green. (1998), *Assessing the accuracy of remotely sensed data: principles and practices*, CRC Press, Lewis, FL, USA, 160p.

Foody, G. M. and P. M. Atkinson. (2002), *Uncertainty in Remote Sensing and GIS*, Wiley & Sons, Hoboken, NJ, USA, 307p.

Kuzera, K. and R.G. Pontius Jr. (2004), "Categorical coefficients for assessing soft-classified maps at multiple resolutions". *In:* H.T. Mowrer, R. McRoberts and P.C. Van Deusen (eds.) *Proceedings of the joint meeting of the 15th annual conference of The International Environmetrics Society and the 6th annual symposium on Spatial Accuracy Assessment in the Natural Resources and Environmental Sciences*, 28 June – 1 July 2004, Portland, ME.

Lewis, H.G. and M. Brown. (2001), "A generalized confusion matrix for assessing area estimates from remotely sensed data". *International Journal of Remote Sensing*, Vol. 22(16): 3223-3235.

Monserud, R.A. and R. Leemans. (1992), "Comparing global vegetation maps with the Kappa statistic". *Ecological Modelling*, Vol. 62: 275-293.

Pijanowski, B.C., K.T. Alexandridis and D. Muller. (2006), "Modeling urbanization patterns in two diverse regions of the world". *Journal of Land Use Science*, Vol. 1(2-4): 83-109.

Pijanowski, B.C., D.G. Brown, B.A. Shellito and G.A. Manik. (2002), "Using neural networks and GIS to forecast land use changes: a land transformation model". *Computers, Environment and Urban Systems*, Vol. 26(6): 553-575.

Pijanowski, B.C., S. Pithadia, B.A. Shellito and K. Alexandridis. (2005), "Calibrating a neural network-based urban change model for two metropolitan areas of Upper Midwest of the United States". *International Journal of Geographical Information Sciences*, Vol. 19: 197-215.

Pontius Jr., R. (2002), "Statistical methods to partition effects of quantity and location during comparison of categorical maps at multiple resolutions". *Photogrammetric Engineering and Remote Sensing*, Vol. 68: 1041-1049.

Pontius Jr., R.G. and M.L. Cheuk. (2006)," A generalized cross-tabulation matrix to compare soft-classified maps at multiple resolutions". *International Journal of Geographical Information Science*, Vol. 20(1): 1-30.

Pontius Jr., R.G. and J. Connors. (2006), "Expanding the conceptual, mathematical and practical methods for map comparison". *In:* M. Caetano and M. Painho (eds.) *Accuracy 2006: 7th International Symposium on Spatial Accuracy Assessment in Natural Resources and Environmental Sciences*, pp. 64-79.

Pontius Jr., R.G., D. Huffaker. and K. Denman. (2004a), "Useful techniques of validation for spatially explicit land-change models". *Ecological Modeling*, Vol. 179(4): 445-461.

Pontius Jr., R.G. and L. Schneider. (2001), "Land-use change model validation by a ROC method". *Agriculture, Ecosystems, and Environment*, Vol. 85: 239-48.

Pontius Jr., R.G., E. Shusas. and M. McEachren. (2004b), "Detecting important categorical land changes while accounting for persistence". *Agriculture, Ecosystems & Environment*, Vol. 101(2-3): 251-68.

Pontius Jr., R.G. and J. Spencer. (2005), "Uncertainty in extrapolations of predictive land change models". *Environment and Planning B*, Vol. 32: 211-230.

SPSS INC. (2003), *SPSS for Windows* (Version 12.0). Chicago, SPSS Inc.

Tayyebi, A., M.R. Delavar, S. Saeedi and J. Amini. (2008a), "Monitoring the urban expansion by multitemporal GIS Maps". *In: Proceedings of the FIG Working Week – Integrating Generations, Stockholm, Sweden.* http://www.fig.net/pub/fig2008/.

Tayyebi, A., M.R. Delavar, S. Saeedi, J. Amini and H. Alinia. (2008b), "Monitoring land use change by multi-temporal LANDSAT remote sensing imagery". *In: Proceedings of the ISPRS Commission VII, The International Society for Photogrammetry and Remote Sensing, Beijing, China*, pp. 1037-1042.

Tayyebi, A., M.R. Delavar, B.C. Pijanowski and M.J. Yazdanpanah. "Urban Expansion Simulation using of Geospatial Information System and Artificial Neural Networks". *International Journal of Environmental Research* (in press).

Walker, R. (2003), "Evaluating the performance of spatially explicit models". *Photogrammetric Engineering & Remote Sensing*, Vol. 69:127-1278.

Wilkinson, G.G. (2005), "Results and implications of a study of fifteen years of satellite image classification experiments". *IEEE Transactions on Geoscience and Remote Sensing*, Vol. 43: 433-440.

# 11

## Deciding on the Detail of Soil Survey in Estimating Crop Yield Reduction Due to Groundwater Withdrawal

*Martin Knotters, Henk Vroon, Arie van Kekem & Tom Hoogland*

### Abstract

*Dutch farmers in areas with groundwater withdrawal receive a yearly legal compensation for crop yield reduction caused by lowered water tables. The crop yield reduction depends on the magnitude of lowering and on soil physical properties. Soil maps are used to estimate crop yield reductions from which annual compensations for farmers are calculated. The risk that farmers receive compensations that are too small can be reduced by paying them 'over-compensation' or by investing in more detailed soil survey. The analysis shows that a detailed soil survey can be an economically attractive alternative to over-compensation. However, to decide on the required map scale, more information on the spatial distribution of classification errors is needed.*

### 11.1 Introduction

In large parts of the Netherlands crop growth depends on the water table, because of its shallow depth (Knotters, 2001; Finke *et al.*, 2004). If groundwater is withdrawn, for instance for drinking water, agricultural crop production might be reduced because of the lowered water tables in the vicinity of the pumping-station. Farmers in these areas receive an annual legal compensation for these losses. The damage to be compensated depends on i) the magnitude of the lowering of water table depth, which decreases with the distance to the pumping-station, and ii) the soil's physical properties. Compensations are calculated on the basis of so-called TCGB-tables, which give percentages of crop yield reduction for combinations of soil types and water table depths (De Laat, 1980; Bouwmans, 1990). In the Netherlands, spatial information on both water table depth and soil type is obtained by soil surveys (Van Heesen, 1970).

Until now, quantitative information on the accuracy of soil maps is not applied in the Netherlands when deciding on the level of legal compensation to farmers in areas with groundwater withdrawal. In fact, the risks of wrong decisions on compensations are unknown, and they appear only as soon as farmers complain about the low level of compensation.

The accuracy of soil maps at various scales has frequently been subject to study, e.g. Marsman and De Gruijter (1986) and Salehi *et al.* (2003). In several studies the consequences of the quality of soil maps of various scales for applications in agronomy and hydrology is analysed. Leenhardt *et al.* (1994) evaluated the efficacy of soil maps of various scales for predicting values of soil water properties at unvisited sites. Lagacherie *et al.* (2000) discussed the accuracy of crop yield estimates based on qualitative soil information in the perspective of decision making. Romanowicz *et al.* (2005) investigated the sensitivity of a rainfall-runoff model to soil input data derived from maps of various scales. Webb and Lilburne (2005) analysed the consequences of soil map unit uncertainty on the assessment of risk of nitrate leaching. Brus *et al.* (1992) discussed upgrading vs. updating of soil maps with respect to the application of soil information in estimating phosphate sorption characteristics. In upgrading statistical information on means and variances is collected for the map units of the existing map, while in updating the map is revised. Basically, the difference between upgrading and updating is analogous to quantifying and reducing uncertainty. This difference is relevant in the decision on the detail of soil information needed for calculating compensations for crop yield reduction.

Errors in calculated compensations for farmers can be reduced by investing in spatial data quality, i.e. more detailed soil survey. However, the budget for these surveys is limited, and the costs should be in good balance with the risks of wrong decisions to be prevented. As an alternative to investing in detailed soil survey, extra compensation can be paid to the farmers to avoid the risk that their losses are not completely compensated, i.e. overcompensation.

A decision has to be made whether uncertainty about spatial soil patterns should be reduced or extra compensation should be paid to farmers to account for errors in soil maps. To support this decision, quantitative information on the accuracy of soil maps is needed. The decision problem can be approached in several ways. We will perform a simple cost-benefit analysis, since in our case both costs and benefits can easily be expressed in monetary values (Baron, 2000). The decisions on quantifying and reducing uncertainty can be ordered in an event-decision tree (Freeze *et al.*, 1992; Kleindorfer *et al.*, 1993). It will be discussed that such an analysis would benefit from knowledge about the spatial structure of classification errors.

The aim of this study is to support the decision on either investing in the quality of spatial data needed for estimating crop yield reduction or in paying extra compensations to farmers to account for errors in spatial data, by a cost-benefit analysis. To this point the costs of detailed soil survey will be compared with the risks of wrong decisions on legal compensation of crop yield reduction. The analysis is illustrated with a case study.

## 11.2  Deciding on the Detail of Soil Survey

### 11.2.1  Uncertainty in estimating crop yield reduction

The calculated crop yield reductions generally will not reflect reality because of errors in several inputs, see Figure 11.1:

1. Errors in soil physical conditions. Soil physical properties vary in space. They are an important factor in the relationship between water table depth and crop production.
2. Errors in information on the water table depth before the withdrawal started.
3. Errors in information on the water table depth since the withdrawal.
4. Errors in the relationship between water table depth, soil type and crop production.

Uncertainty about crop yield reduction due to errors in 1, 2 and 3 can be reduced by making more detailed soil maps. Note that in the Netherlands the water table depth is mapped concurrently with the soil (Van Heesen, 1970). Errors in 4 cannot easily be reduced. The relationship between water table depth, soil type and crop production, summarized in the TCGB-tables, reflects years of agro-hydrological research and will be considered as 'true' in this study (Bouwmans, 1990). We will focus on reduction of the first three sources of uncertainty.

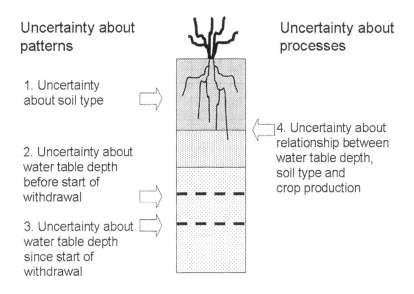

**Figure 11.1.** Sources of uncertainty in calculating crop yield reduction due to groundwater withdrawal.

### 11.2.2 Analysis of costs and risks

Due to spatial variation of hydrological conditions and soil properties the compensations paid to the farmers vary in space. Inaccurate information on spatial patterns may result in compensations that are too large or too small. The risk that compensations are too small can be reduced in two ways. The first way is to 'over-compensate' farmers. The second way is to invest in a detailed inventory of soil and hydrological patterns. We will choose between these two options by comparing the costs of over-compensation with the costs of a detailed survey in a cost-benefit analysis. Figure 11.2 illustrates the cost-benefit analysis.

To compare the costs of soil survey with the risks of wrong annual compensations paid to farmers, the magnitude of errors need to be estimated and the net present values of errors in yearly compensations need to be calculated.

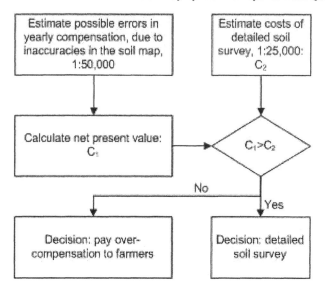

**Figure 11.2.** Flowchart of the cost-benefit analysis.

We estimated the magnitude of errors by selecting 25 combinations of soil type and water table depth randomly from the TCGB-tables. We considered the related percentages of crop yield reduction as being the 'true' situation before the ground-water withdrawal started. Next we selected for each of these 25 'true' situations a misclassification. This selection was performed randomly from a uniform distribution, allowing for 0-5 classes deviation for the type of subsoil, and for 0-3 classes deviation for the soil water retention curve of the root zone, the mean spring water table depth, the mean summer (or lowest) water table depth and the thickness of the root zone. In this way unrealistic misclassifications were prevented. Finally we randomly selected 25 numbers from a uniform distribution, assigned to the lower-ing of water table depth between 50 and 5cm following an exponential shape which approximates the shape of the water table around a groundwater extraction site. The selected 25 lowerings were applied to both the 25 'true' situations and the 25 mis-classifications. The differences between crop yield reduction for the 'true' situa-tions and the misclassifications indicate the magnitude of error due to inaccuracies in the soil map.

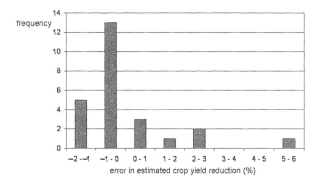

**Figure 11.3.** Distribution of errors in crop yield reduction, due to classification errors in the soil map.

Figure 11.3 shows the error distribution. In the worst case a farm is completely situated in an impure part of the soil map, in which the crop yield reduction is underestimated by 6%. In this case the compensation paid to the farmer will be too small. Given the surface area of a farm, the risk of underestimated crop yield reduction depends on i) the scale of the soil map and ii) the spatial distribution of classification errors. For the national 1:50,000 soil map the risk that crop yield reduction of a farm is underestimated by 6% is larger than for a 1:25,000 soil map. If classification errors show strong variation at a short distance, then the risk that a farmer will not receive enough compensation will also be low: negative and positive errors sum to approximately zero.

If detailed soil maps are not available and the spatial distribution of classification errors is unknown, farmers should be overcompensated to avoid an underestimation of 6% crop yield reduction. This overcompensation equals €70.80 yearly per hectare (€11.80 per percent reduction, per hectare). The net present value of this overcompensation is calculated by:

$$C = b_t \frac{1-(1+r)^{-T}}{r} \, ,$$

with $b_t$= €70.80, $r$ is the interest, say, 6 %, and $T$ is the planning horizon, say, 30 years. This results in a net present value of €975 over-compensation per hectare. From recent surveys we know that the costs of a detailed soil survey are much lower: a 1:25,000 soil map, with 1.80-2.50m augering depth, costs €55 per hectare. A detailed soil survey is attractive in this case, provided that the spatial structure of errors is such that they sum to zero within the surface area of a farm.

## 11.3  Case Study

We estimated crop yield reductions using the national 1:50,000 soil map and a 1:25,000 soil map, for four farms in the vicinity of a pumping-station in the south-eastern part of the Netherlands. The four farms are situated in a zone where the water table has been lowered by 30cm since the groundwater withdrawal started. Figure 11.4 shows the patterns of the estimated crop yield reductions within the four farms. As compared to the 1:50,000 map, the 1:25,000 soil map gives more detailed information about spatial patterns and the structure of the subsoil. Table 11.1 gives the annual compensations, calculated for the four farms on the basis of the 1:50,000 and the 1:25,000 map. The largest difference in yearly compensation is found for farm C: zero compensation vs. a compensation of €49 $ha^{-1} \cdot y^{-1}$ (net present value €674 $ha^{-1}$). Thus, an investment of €55 $ha^{-1}$ in a detailed soil survey results in a considerable increase of compensation for farm C: from €0 to €674 $ha^{-1}$.

Farm B would receive €6 $ha^{-1} \cdot y^{-1}$ less compensation if the calculations were based on the 1:25,000 soil map instead of the national soil map, 1:50,000 (net present value €93 $ha^{-1}$).

For the total area of the four farms we calculated the absolute differences in compensations and compared these with the costs of the detailed soil survey: the net present value of the absolute differences is €212 $ha^{-1}$, which is considerably more than the costs of the soil map, 1:25,000 (€55 $ha^{-1}$). This indicates that investing in quality of spatial data on soil is justified from an economic point of view.

**Table 11.1.** Estimated crop yield reductions and compensations, based on a 1:25,000 soil map, and the national 1:50,000 soil map. A, B, C and D indicate the farms for which annual compensations are calculated (Figure 11.4).

| Farm | National soil map, 1:50,000 | | Soil map, 1:25,000 | |
|------|------------------------------|--|---------------------|--|
|  | Crop yield reduction | Compensation | Crop yield reduction | Compensation |
|  | (%) | (€·$ha^{-1} \cdot y^{-1}$) | (%) | (€·$ha^{-1} \cdot y^{-1}$) |
| A | 5.5 | 64 | 5.7 | 67 |
| B | 5.2 | 61 | 4.6 | 55 |
| C | 0 | 0 | 4.1 | 49 |
| D | 3.5 | 41 | 3.7 | 44 |

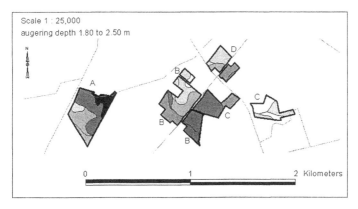

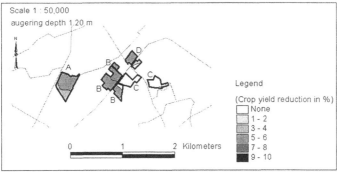

**Figure 11.4.** Estimated crop yield reductions, based on a 1:25,000 soil map (top), and the 1:50,000 national soil map (bottom). A, B, C and D indicate the farms for which annual compensations are calculated (Table 11.1).

## 11.4 Discussion and Conclusion

The result of the analysis indicates that detailed soil maps can be valuable in calculating legal annual compensations to farmers in areas influenced by groundwater withdrawal. However, more information on the spatial distribution of classification errors is needed to choose between map scales. Figure 11.5 shows notional examples of four situations which can occur.

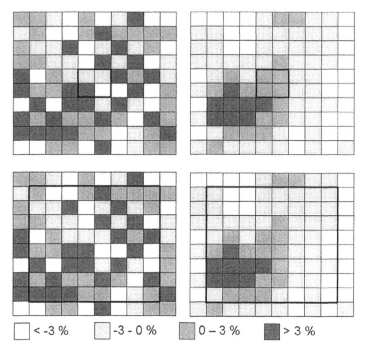

☐ < -3 %    ☐ -3 - 0 %    ▨ 0 – 3 %    ▦ > 3 %

**Figure 11.5.** Notional examples of underestimation of crop yield reductions, due to classification errors at the soil map. The bold frame indicates a farm area. Top left: small map scale, no spatial structure in classification errors. Top right: small map scale, strong spatial structure in classification errors. Bottom left: large map scale, no spatial structure in classification errors. Bottom right: large map scale, strong spatial structure in classification errors.

If the map scale is small relative to the surface area of a farm (top of Figure 11.5), there is a considerable risk that crop yield reduction of a farm is underestimated or overestimated. By moving the position of the farm area over the map pixel-by-pixel, the under- or overestimation of crop yield reduction varies. The risk of under- or overestimation increases with increasing spatial structure of the errors (compare top left with top right in Figure 11.5). If the map scale is large relative to the surface area of a farm, and there is no spatial structure in errors (bottom left in Figure 11.5), the errors will approximately sum to zero within the area of a farm. In this situation a farmer has little risk of underestimated crop yield reduction. In the case of strong spatial structure of errors there is a risk of underestimation or overestimation (bottom right in Figure 11.5).

In summary, given the large costs of 'over-compensation' it is attractive to invest in detailed soil survey. However, more information on the spatial distribution of classification errors at soil maps is needed to decide on the detail of soil survey required for the accurate estimation of crop yield reduction in areas with groundwater withdrawal. An interesting next step will be to estimate the risks of wrong decisions on compensations paid to farmers, for various map scales and various spatial structures of errors, in a real case, on the basis of an independent validation set. A decision on the required map scale can be supported by the results of such an analysis.

# References

Baron, J. (2000), *Thinking and deciding*. Cambridge University Press, Cambridge, 570p.

Bouwmans, J.M.M. (1990), *Background and application of the TCGB-table. A method to calculate crop yield reduction of grassland at a sandy soil due to lowering of the water table depth* (in Dutch). TCGB, Utrecht, 189p.

Brus, D.J., J.J. de Gruijter and A. Breeuwsma. (1992), "Strategies for updating soil survey information: a case study to estimate phosphate sorption characteristics". *Journal of Soil Science*, Vol. 43: 567-581.

De Laat, P.J.M. (1980), *Model for unsaturated flow above a shallow water table, applied to a regional sub-surface problem*. PhD thesis, Wageningen University, The Netherlands, 126p.

Finke, P.A., D.J. Brus, M.F.P. Bierkens, T. Hoogland, M. Knotters and F. de Vries. (2004), Mapping groundwater dynamics using multiple resources of exhaustive high resolution data. *Geoderma*, Vol. 123: 23-39.

Freeze, R.A., B. James, J. Massmann, T. Sperling and L. Smith. (1992), "Hydrogeological decision analysis: 4. The concept of data worth and its use in the development of site investigation strategies". *Ground Water*, Vol. 30: 574-588.

Kleindorfer, P.R., H.C. Kunreuther and P.J.H. Shoemaker. (1993), *Decision sciences: an integrative perspective*. Cambridge University Press, Cambridge, 470p.

Knotters, M. (2001), *Regionalised time series models for water table depths*. PhD thesis, Wageningen University, The Netherlands, 153p.

Lagacherie, Ph., D.R. Cazemier, R. Martin-Clouaire and T. Wassenaar. (2000), "A spatial approach using imprecise soil data for modelling crop yields over vast areas". *Agriculture, Ecosystems and Environment,* Vol. 81: 5-16.

Marsman, B.A. and J.J. de Gruijter. (1986), *Quality of soil maps. A comparison of soil survey methods in a sandy area*. Soil Survey Papers No. 15, Wageningen, The Netherlands, 103p.

Romanowicz, A.A., M. Vanclooster, M. Rounsevell and I. La Junesse. (2005), "Sensitivity of the SWAT model to the soil and land use data parametrisation: a case study in the Thyle catchment, Belgium". *Ecological Modelling,* Vol. 187: 27-39.

Salehi, M.H., M.K. Eghbal and H. Khademi. (2003), "Comparison of soil variability in a detailed and a reconnaissance soil map in central Iran". *Geoderma,* Vol. 111: 45-56.

Van Heesen, H.C. (1970), "Presentation of seasonal fluctuation of the water table on soil maps". *Geoderma*, Vol. 4: 257-278.

Webb, T.H. and L.R. Lilburne. (2005), "Consequences of soil map unit uncertainty on environmental risk assessment". *Australian Journal of Soil Research,* Vol. 43: 119-126.

# 12
## Short Notes

## A New Method for Integrating Data of Variable Quality into 3-Dimensional Subsurface Models

*Kelsey MacCormack & Carolyn H. Eyles*

Three-dimensional (3D) models are currently being used by both private industry and government agencies to visualize subsurface geological characteristics as they readily communicate complex concepts to both specialists and the general public (Kessler *et al.*, 2005; Jackson, 2007). These 3D models can be used for a variety of applications including resource exploration purposes (Paulen *et al.*, 2006), environmental applications such as groundwater source protection (Burt, 2007), and to aid in reconstruction of past geologic events and processes (Mac-Cormack *et al.*, 2005; Logan *et al.*, 2006). The data necessary to generate these models are often difficult to obtain and typically come from sources such as regional waterwell databases (e.g. the Ontario Waterwell Database), engineering and construction records, and geological survey reports. The recent high demand for 3D subsurface models has resulted in an increased dependency on the large, digital waterwell databases in order to produce models in relatively short time frames, because they contain large amounts of subsurface information with good spatial distribution. However, a serious concern with utilizing large digital databases, such as the waterwell database, as the primary (or only) source of data is the variable quality and reliability of the input data (Goodchild and Clark, 2002; Logan *et al.*, 2006; Burt, 2007; Venertis, 2007). Three-dimensional models are predictions based solely on the data provided and therefore the accuracy and reliability of model outputs are severely constrained by the quality of input data. This study focuses on the development of a new method for integrating data of varying quality into the modeling process while allowing the high quality data to take precedence in the spatial interpolation of subsurface geologic units. Subsurface geologic data available from the McMaster University campus in Hamilton, Ontario are used to illustrate how significantly the quality of input data can impact the 3D model outputs, and the effectiveness of this new method of data integration.

Subsurface data for an approximately 0.8km$^2$ area of the McMaster University campus were obtained from construction and engineering reports, the Hamilton urban geology database, and the Ontario Waterwell Database. The quality of individual borehole records was determined based on the level of detail in the accompanying sediment descriptions and also on the degree of correlation of subsurface units between adjacent wells. The 'higher quality' borehole records were obtained from engineering and construction reports in which the depth and characteristics of the subsurface sediment units were the focus of the investigation at the time of drilling. The 'lower quality' borehole records typically came from the large digital databases which often provided general sediment classifications with little or no accompanying description, and which often did not correlate well with neighbouring borehole logs. The digital records available from regional waterwell databases are based on logs generated by drillers whose focus is on finding water rather than on accurately logging the subsurface sediments, and the accuracy of the sediment descriptions are often questionable. In an ideal situation, only high quality data would be used to create 3D subsurface models. However these data are rarely sufficient to provide adequate spatial coverage for modelling purposes, making it necessary to include some lower quality data in order to constrain the model where data are sparse.

Current 3D geologic modelling software programs are only able to consider the value of the variable of interest at a given data point and do not include a means to incorporate information on the reliability or accuracy of the value itself. As the 3D modelling software assigns an equal weighting to both high and low quality data in the modelling process, this ultimately causes a significant 'dilution' of the high quality data and compromises the reliability of the model output. In order to visualize the effects of data quality on 3D model output, high and low quality data points identified from the McMaster campus study area were subdivided into two datasets and modelled separately using RockWorks 2006. Each of the grids created by the high and low quality datasets was then multiplied by a 'weighting factor' to either enhance or decrease the influence of the respective data on the final output model. The weighted grids were then mathematically recombined with one another using a grid math process. Assigning a larger relative weighting factor to the grid created by the high quality data allows it to have a greater influence on the final model output. This methodology was also shown to improve the model output in areas where the high quality data were sparse, by allowing the lower quality data to help constrain the model output in these areas. A range of relative 'weighting' values were tested on both the high and low quality datasets and it was found that the recombined weighted model of 70% high quality and 30% low quality data produced an output model that most closely honoured the high quality data, while utilizing the lower quality data to fill in the areas of sparse data coverage. This recombined weighted model also showed a significant improvement over the non-weighted (equal weighting of high and low data) model output with respect to honouring the high quality data values.

The effectiveness of this method of integrating data of different quality into the modelling process was also assessed by quantifying the differences between the output models created from the low quality, high quality, and recombined weighted datasets by calculating the predicted volume for one of the modelled subsurface units. In this study, a coarse sandy gravel subsurface unit was chosen as it forms a significant local aquifer. This aquifer was the focus of an independent contaminant migration study that provided proxy data on aquifer characteristics with which to compare the modelled results. Calculations of the aquifer volume based on each of the modelled datasets showed considerable variation with the low quality dataset estimating values up to 79% greater than those produced by the high quality dataset. These results show that very significant uncertainties may be introduced into the model predictions through the use of low quality or highly variable quality data.

This study demonstrates that data quality is an important factor to be considered when assessing the reliability of 3D subsurface model outputs. A methodology is presented here that allows high quality data to be more influential in constraining the interpolation of unit boundaries in localized areas during the modeling process, while utilizing lower quality data to expand the model regionally where data are sparse. This 'integrated' model approach utilizes the strengths of both data sources and can readily be applied to enhance the quality of 3D subsurface model outputs in other regions.

# References

Burt, A.K. (2007), "Three-dimensional geological modelling of thick Quaternary deposits in the Barrie-Oro area, central Ontario: new modelling techniques". *In* C.L. Baker, E.J. De-bicki, J.R. Parker, J.K. Mason, R.I. Kelly, M.C. Smyk, J.A. Ayer, G.M. Stott and P. Sarvas (eds.). *Summary of Field Work and Other Activities 2007, Ontario Geological Survey, Open File Report 6213*, pp. 21-1 - 21-9.

Goodchild, M.F., and K.C. Clarke. (2002), "Data quality in massive data sets". *In:* J. Abello, P.M. Pardalos and M.G.C. Resende (eds.). *Handbook of massive data sets*, Kluwer Academic Publishers, Norwell, USA, pp. 643-659.

Jackson, I. (2007), "So where do we go from here?". *Minnesota Geological Survey, Open File Report 07-4*, pp. 19-22.

Kessler, H., M. Lelliott, D. Bridge, F. Ford, H.G. Sobisch, S. Mathers, S. Simon Price, J. Merritt, and K. Royse. (2005), "3D geosciences models and their delivery to customers". *Geological Survey of Canada, Open File 5048*, pp. 39-42.

Logan, C., H.A.J. Russell, D.R. Sharpe, and F.M. Kenny. (2006), "The role of GIS and expert knowledge in 3-D modelling, Oak Ridges Moraine, southern Ontario". *In:* J.R. Harris (ed.), *2006, GAC Special Paper 44: GIS for the Earth Sciences, Geological Association of Canada*, St. John's, Canada, pp. 519-541.

MacCormack, K.E., J.C. Maclachlan, and C.H. Eyles. (2005), "Viewing the subsurface in three dimensions: Initial results of modeling the Quaternary sedimentary infill of the Dundas Valley, Hamilton, Ontario". *Geosphere*, Vol. 1(1): 23-31.

Paulen, R.C., M.B. McClenaghan, and J.R. Harris. (2006), "Bedrock topography and drift thickness models from the Timmins area, northeastern Ontario: An application of GIS to the Timmins overburden drillhole database". *In:* J.R. Harris (ed.), *2006, GAC Special Paper 44: GIS for the Earth Sciences, Geological Association of Canada*, St. John's, Canada, pp. 413-434.

Ventris, E.R. (2007), "3-D modeling of glacial stratigraphy using public water well data, geologic interpretation, and geostatistics". *In:* L.H. Thorleifson, R.C. Berg, and H.A.J. Russell (eds.), *Three-dimensional Geologic mapping for Groundwater Applications Extended Abstracts*, Open file report 07-4, pp. 81-84.

# Accuracy Measurement for Area Features in Land Use Inventory

*Haixia Mao, Wenzhong Shi & Yan Tian*

Because of the importance of the land use inventory, it is used in many countries, and results in the creation of map parcels. The accuracy of patches of area features and how to assess their accuracy are always open issues in land use inventory. In this paper, indicators for measuring attribute accuracy of an area are proposed, which cover (a) area similarity, and (b) boundary similarity. The corresponding mathematical formulae for the proposed indicators are given.

## Area similarity

The main data products in a land use inventory are the map parcel and linear features, and so it is required that the area of the feature polygon be described. If there is a feature $i$ in the land use map and a corresponding reference feature $iR$, we can calculate the area similarity between two map parcels to describe their difference:

$$AS(S_i, S_{iR}) = \frac{|S_i - S_{iR}|}{S_{iR}} \tag{12.1}$$

where $S_i$ and $S_{iR}$ is the area of a certain map parcel from the product of a land use inventory and its corresponding reference data respectively. $AS(S_i, S_{iR})$ represents the degree of area similarity between $S_i$ and $S_{iR}$.

## Boundary similarity

Generally, different polygons may have the same area but a different shape. It is the same situation in a land use inventory. Consequently we need to complement area similarity with the use of boundary similarity to measure the attribute accuracy.

(1) Polar coordinate representation.

Rather than the Cartesian coordinate system, we use the polar coordinate system to describe the boundary similarity between two features. The benefit is, in this case, the boundary can be described with a 1-dimension curve.

The gravity of a polygon is always considered the origin in a polar coordinate system, and the range is from the gravity to the boundary. It is then possible to transfer the 2-dimensional shape boundary into the 1-dimension $f(\theta)$ function with the polar coordinate representation:

$$f(\theta) = \frac{r(\theta)}{\max(r(\theta))}, \theta \in [0, 2\pi) \tag{12.2}$$

Here, $r(\theta)$ is the range of the boundary pixel points with $\theta$ angle, $\max(r(\theta))$ is the largest of all the ranges. So, $f(\theta)$ is a normalized shape description curve.

(2) Shape distance and boundary similarity.

Suppose $f(\theta, \alpha)$ is obtained by left rotating $f(\theta)$ with $\alpha$ angle clockwise, we have:

$$f(\theta, \alpha) = \begin{cases} f(\theta + \alpha), \theta + \alpha < 2\pi \\ f(\theta + \alpha - 2\pi), \theta + \alpha \geq 2\pi \end{cases} \tag{12.3}$$

Based on Equation 12.3, if $f(\theta)$ is the curve function of boundary $S$, and $S'$ is obtained by rotating $s$ with $\alpha$ angle clockwise, then the curve function of boundary $S'$ is $f(\theta, \alpha)$.

To different boundary $S_1$ and $S_2$, suppose their curve functions are $f_1(\theta)$ and $f_2(\theta)$ respectively, $\theta \in [0, 2\pi)$, then the shape distance between them can be defined as $D(S_1, S_2)$:

$$D(S_1, S_2) = \min_a \int_0^{2\pi} |f_1(\theta) - f_2(\theta, a)| d\theta \tag{12.4}$$

Suppose boundary $S_2$ rotates $A$ angle clockwise (namely $\alpha = A$ in Equation 12.4), then $\int_0^{2\pi} |f_1(\theta) - f_2(\theta, a)| d\theta$ will obtain the smallest value, and angle $A$ can be defined as the direction difference degree between two boundaries. Equation 12.4 can also be further revised as:

$$D(S_1, S_2) = \int_0^{2\pi} |f_1(\theta) - f_2(\theta, A)| d\theta \tag{12.5}$$

Boundary similarity can be defined based on shape distance, so the boundary similarity between $S_1$ and $S_2$ can be defined as:

$$C(S_1, S_2) = 1 - \frac{\int_0^{2\pi} |f_1(\theta) - f_2(\theta, A)| d\theta}{\int_0^{2\pi} d\theta} = 1 - \frac{D(S_1, S_2)}{2\pi} \tag{12.6}$$

If one selects an appropriate threshold $D_T$ (or $C_T$, $C_T = 1 - \frac{D_T}{2\pi}$), when $D(S_1, S_2) \leq D_T$, or $C(S_1, S_2) \geq C_T$), then we can assume that boundary $S_1$ and $S_2$ belong to the same shape.

The direction difference degree $A$ is a key parameter in Equations 12.5 and 12.6. Correct selection can keep the rotation inflexibility of shape curve representative of the object shapes.

There are some previous works on this matter; however, there is currently no excellent solution for this issue. In this paper, we proposed the area similarity and boundary similarity to quantitatively measure the accuracy of map parcels for land use inventory. Although only general ideas are presented here, deeper and more experimental work will be carried out in the future.

## Acknowledgments

This work was supported by funds from The Hong Kong Polytechnic University (Project No.: RGMG and G-YF24), and Hong Kong RGC (Project No.: 5276/08E).

# Granular Computing Model for Solving Uncertain Classification Problems of Seismic Vulnerability

*Hadis Samadi Alinia & Mahmoud Reza Delavar*

One of the most well-known risks affecting urban areas is the earthquake. Tehran is a megacity located in an earthquake-prone region, based on the historical records available. Therefore, it is predicted that a huge earthquake in Tehran is imminent. The production of a seismic vulnerability map therefore helps local and national disaster management organizations to create and implement a plan to promote awareness of earthquake vulnerability, and assists in the implementation of seismic vulnerability reduction measures in Tehran.

Production of a seismic vulnerability map generally depends on various criteria. One of our challenges is to implement suitable methods for granulation and graduation of the different related factors used to classify urban areas. The problem raised here is the different quality of spatial factors used in the decision making process, and also the existence of uncertainties in these factors.

Many structures and methodologies have been proposed to resolve uncertainties. Each of them provides a set of concrete methods and tools for solving a particular type of problem in a particular context. It has been clearly recognized that the basic ideas of crisp information granulation have appeared in many fields, such as artificial intelligence, interval analysis, quantization, and rough set theory (Yao, 2004a; 2004b; 2005; Chen and Yao, 2008). Human problem-solving involves the perception, abstraction, representation and understanding of real world problems, as well as their solutions, at different levels of granularity (Zadeh, 1997a; Yao, 2000; 2004a; Zhang and Zhang, 2003). The consideration of granularity is motivated by the practical need for simplification, clarity, low cost, approximation and tolerance of uncertainty (Zadeh, 1997b). As an emerging field of study, granular computing attempts to formally investigate and model the family of granule-oriented problem-solving methods and information processing paradigms (Yao, 2000).

To provide more precise vulnerability mapping and absorption and evaluate inherent uncertainties in expert opinions, we propose a granule-centered method in which one granule is defined by an attribute-value pair in each step.

In this research we have used a granular computing model to classify urban areas using an information table in which an urban area is described by a set of attribute-value pairs. Based on this approach, we developed an information table of one hundred urban areas in Tehran to derive the rules. A logic language was defined for the information table to provide formal descriptions of various notions. In order to resolve the attribute-centered strategy, in which unnecessary attributes in classification rules may be introduced, we used granule-centered strategies, where one granule was defined by an attribute-value pair in each step. After dividing the universe into subsets or grouping individual objects into clusters, granules that were subsets of the universe were achieved. Then, to identify the uncertainty of an urban area, we determined conditional entropy. Finally, based on measures in previous steps, a consistent classification problem defined the granule-based rule induction method proposed.

This paper outlines the potential of granular computing to reduce the uncertainty classification of seismic physical vulnerability urban area into six classes in which objects (urban blocks) are described by a finite set of attributes assuming that the north fault of Tehran is activated: these are earthquake intensity (MMI unit); mean slope; weak buildings with fewer than five floors; weak buildings with five or more floors; building structure (well built or not reinforced, well built and reinforced); and age of building.

The focus of this paper is to induce some rules and a subset of attributes from information table by granular network to classify urban block with less uncertainty in seismic vulnerability modeling. We will show that granular computing can not only make a complex problem more easily understandable, but also leads to efficient solutions.

## References

Chen, Y.H. and Y.Y. Yao. (2008), "A multiview approach for intelligent data analysis based on data operators". *Information Sciences*, Vol. 178(1): 1-20.

Yao, Y.Y. (2000), "Granular computing: basic issues and possible solutions". *In:* P.P. Wang (ed.) *Proceedings of the 5th Joint Conference on Information Sciences, Volume 1*, Association for Intelligent Machinery, Atlantic City, USA, pp. 186-189.

Yao, Y.Y. (2004a), "A partition model of granular computing". *LNCS Transactions on Rough Sets*, Vol. 1: 232-253 (LNCS 3100).

Yao, Y.Y. (2004b), "Granular computing". *Proceedings of the 4th Chinese National Conference on Rough Sets and Soft Computing, Computer Science (Ji Suan Ji Ke Xue),* Vol. 31: 1-5.

Zadeh, L.A. (1997a) "Towards a theory of fuzzy information granulation and its centrality in human reasoning and fuzzy logic". *Fuzzy Sets and Systems*, Vol. 19: 111-127.

Zadeh, L.A. (1997b), *Some reflections on soft computing, granular computing and their roles in the conception, design and utilization of information/intelligent systems*, Computer Science Division, University of California, Berkeley, USA.

Zhang, L. and Zhang, B. (2003) "The quotient space theory of problem solving". *In:* G. Wang, Q. Liu, Y. Yao, and A. Skowron (eds.) *Rough Sets, Fuzzy Sets, Data Mining, and Granular Computing.* LNCS 2639, pp. 11-15.

# Spatial Variability of Errors in Urban Expansion Model: Implications for Error Propagation

*Amin Tayyebi, Mahmoud Reza Delavar, Bryan C. Pijanowski &
Mohammad J. Yazdanpanah*

Key decisions in areas as diverse as environmental monitoring and town planning are made using a range of GIS modeling techniques. Many of these involve the combination of multiple raster data input layers, all of which contain some degree of uncertainty. That uncertainty may vary spatially throughout a dataset. The way in which uncertainty is propagated through a GIS map algebra function depends on the mathematical shape of that function. It can be shown (Burrough and McDonnell, 1998), that some GIS operations are more prone to exaggerate error than others, with exponentiation being particularly vulnerable. This is, however, also dependant on the magnitude of the input values. The Urban Expansion Model (UEM) is adopted through utilizing Geospatial Information System (GIS), Artificial Neural Networks (ANNs) and Remote Sensing (RS) for simulation of urban expansion (Tayyebi *et al.*, in press). The UEM follows eight sequential steps including: (1) rectification and registration; (2) classification; (3) integrating topographic data in a database; (4) coding of data to create spatial layers of predictor variables; (5) applying spatial or non-spatial function in ArcGIS; (6) integrating all input grids; (7) calibration of UEM; and (8) temporal prediction (Tayyebi *et al.*, in press).

This paper examines two different methods by which error is propagated through an UEM. The two methods are, firstly, an analytical method, referred to here as a first order Taylor series method, and secondly the Monte Carlo simulation method (Marinelli *et al.*, 2006). These are assessed to determine how the error propagation results vary between models. Ideally they should also be compared from the standpoints of the ease with which these methods can be used, and the time required for the actual computation of the error analysis. That work is beyond the scope of this paper, but is the subject of ongoing research. The method of error propagation usually referred to as the Taylor series method, relies on using either the first or second differences of the function under investigation. In this work the First Order method was used. If we consider a general model with a number of spatial data layers $A_1$ to $A_n$ as input, then any individual cell in the calculation may be represented as in Equation 12.7 (Heuvelink, 1998).

$$u = f\left(a_1, a_2, \ldots a_n\right) \tag{12.7}$$

where: $u$ is the cell value for the model output, and $a_i$ are the individual cell values for each input.

In many applications, particularly where the data layers have been derived by interpolation from multiple observations, the standard deviation is used as a surrogate for the error term. Care should be taken that this is not the standard deviation of the input data layer itself, since a very precise data layer with high natural variability will have a high standard deviation which will not be representative of the accuracy of the data layer. If correlation exists, then it can be shown in Equation 12.8.

$$Eu_\rho = \sum_{j=1}^{n}\sum_{i=1}^{n}\left(\delta u/\delta a_i\right)\times\left(\delta u/\delta a_j\right)\times Ea_i \times Ea_j \times \rho_{ij} \qquad (12.8)$$

where $Eu_\rho$ is additional absolute error in the output; $Ea_i$, $Ea_j$ are the absolute error in the inputs; $\delta u/\delta a_i$, $\delta u/\delta a_j$ are the partial differentials of the function with respect $a_i$,

and $a_j$; $\rho_{ij}$ is the correlation between $a_i$ and $a_j$.

The Monte Carlo method of error propagation assumes that for any GIS function, such as that shown in Equation 12.7, the distribution of error for each of the input data layers $A_i$ is known. The distribution is frequently assumed to be Gaussian, but other distributions may be used. For each of the data layers, an error surface is simulated by drawing, at random, from an error pool defined by this distribution. Those error surfaces are added to the input data layers and the model is run using the resulting data + error layers as input. The process is repeated a large number of times so that, for each run, a new realization of an error surface is generated for each input data layer. The results of each run are accumulated and a running mean and standard deviation surface for the output is calculated. This process continues until the running mean stabilizes. Since the random error visualizations are both positive and negative, the stable running mean may be taken as the true model output surface and the standard deviation surface may be used as a measure of relative error.

Good agreement is found between the spatial distributions of the resulting error surfaces. However, the magnitude of the error calculated using the full analytical method that incorporates a term for the correlation between input data sets is larger than expected. The model is then examined for sensitivity to its inputs and found to be more sensitive to one of them. Some limiting conditions are also noted.

## References

Burrough, P.A. and R.A. McDonnell. (1998), *Principles of Geographical Information Systems*, Oxford, Oxford University Press, 333p.

Marinelli, M., R. Corner and G. Wright. (2006), "A comparison of error propagation analysis techniques applied to agricultural models". *In:* M. Caetano and M. Painho (eds.) *Accuracy 2006. 7<sup>th</sup> International Symposium on Spatial Accuracy Assessment in Natural Resources and Environmental Sciences*,

http://www.spatial-accuracy.org/2006/PDF/Marinelli2006accuracy.pdf

Heuvelink, G.B.M. (1998), *Error Propagation in Environmental Modelling with GIS*, Taylor and Francis, London, 127p.

Tayyebi, A., M.R. Delavar, B.C. Pijanowski and M.J. Yazdanpanah. (in press), "Urban Expansion Simulation using of Geospatial Information System and Artificial Neural Networks". *International Journal of Environmental Research*.

# SECTION III – IMAGERY

# 13

# A Stereological Estimator for the Area of Flooded Land from Remote Sensing Images

*Alfred Stein, Mamushet Zewuge Yifru & Petra Budde*

## Abstract

*This paper introduces the use of stereology in the field of image mining. Image mining considers the chain from object identification from remote sensing images to communication to stakeholders. Stereology quantifies objects at one dimension from simulated objects at a lower dimension. It is presented as a method for measuring the size of spatial objects identified on images. The paper is illustrated with a case study from Cambodia.*

## 13.1 Introduction

Remote sensing images are becoming available at an increasing frequency and spatial resolution, and from a large variety of sensors. Spatial resolution of civil satellites is down to below 1m for the panchromatic bands of the Quickbird and Ikonos satellites, whereas the temporal resolution is down to 15 minutes, e.g. from the Meteosat 2nd generation satellite MSG-1 and MSG-2, although at spatial resolutions of 1 to 5km. Despite this development, uncertainty is still largely present.

We increasingly see it as an important step to summarize the information in a quantitative way. In the recent past, the main focus was on measuring the size of an object including its uncertainty (Shi, 2009; Stein *et al.*, 2008).

Image mining focuses on extracting relevant information from large sets of remote sensing images (Stein, 2008; Silva *et al.*, 2008). It can be defined as *"[the] analysis of (often large sets of) observational images to find (un)suspected relationships and to summarize the data in novel ways that are both understandable and useful to stakeholders"*. Image mining is defined both in space, with a focus on combining many images, and in space × time, with a focus on tracking objects. In space it concerns classification and segmentation, for example using textural image segmentation as a first step. Its main objective is to reduce uncertainty, allowing better decision-making (Frank, 2008). Important steps during image mining are: the identification of spatial patterns and testing of their significance; dimension reduction to have an improved informative content; making realistic assumptions to optimally use existing information; inclusion of all determining factors and making quantitative statements with associated uncertainties. These are related to issues of data quality that also depend upon the fitness for use, i.e. the required decision and hence on the stakeholders' interests.

In this study we aim to introduce stereology in remote sensing, and for this study we will focus on estimation of the size of 2D objects. Stereology is an unbiased and effective tool to obtain quantitative 2D geometric properties (e.g. number, length, area, volume, etc.) from a recorded series of sections in a lower dimension. Quite generally, stereology can be stated as a problem of measuring a physical object in $n$ dimensions from random measured objects in fewer dimensions (Baddeley, 1991; Baddeley and Vedel Jensen, 2005). Stereology, as far as we know, has not previously been applied towards remote sensing images.

The aim of this paper is thus to present essentials of stereology towards analysis for a single remote sensing image. Special attention will be given to issues of data quality. The study will be illustrated with an example from flooding in Cambodia.

## 13.2  Image Mining

In image mining discernable objects have to be identified. We distinguish five key steps in image mining: identification, modeling, tracking, prediction and communication with stakeholders. All these steps will be briefly discussed below. On top of this, we observe aspects of spatial data quality in each of these steps.

Step 1. Object identification. We make the step from raster to objects by grouping grid cells with similar digital values into one object. This is usually done by applying an image segmentation technique (Lucieer *et al.*, 2005). Segmentation is followed by a classification into a set of object classes, defined implicitly or explicitly beforehand (Richards and Jia, 1998). Modeling of uncertainty has been done in the past by using, e.g. a confusion index, or posterior probabilities to the non-selected classes. The result of this operation for an image is a series of $n$ objects $X_i$, $i = 1,\ldots,n$, that belong to a set of $K$ classes $C_k$, $k = 1,\ldots,K$.

Step 2. Modeling of the identified objects. We distinguish fuzzy modeling (Zadeh, 1965; Dilo *et al.*, 2007), random modeling (Stoyan and Stoyan, 1994) and mathematical modeling. Both the fuzzy and random approaches offer a logical frame for handling uncertainty, using the confusion index or the posterior probability, respectively. Mathematical modeling concerns the use of differential equations or linear equations and as such poses more problems in handling uncertainty.

Step 3. Tracking. Tracking objects in space and time is relatively straightforward for an object characterized by a limited set of parameters. It requires, though, a proper modeling of both splitting and merging. In addition, tracking requires one to carefully consider the birth and death of objects.

Step 4. Prediction. Prediction of objects in space at an unobserved time $t_0$ can be done by a parametric curve for the centroid and other parameters. Rajasekar *et al.* (2006) showed the use of a linear statistical model. Prediction considers either a future event, i.e. a real prediction, or a past event or an event between two observation times $t_h$ and $t_{h+1}$. Predicted values are typically of interest to stakeholders, who may thus have a tool to support their decision-making.

Step 5. Communication to stakeholders. Various methods exist here, ranging from simple visualization towards assessments of costs and benefits. Issues from decision support typically are required here (Van de Vlag and Stein, 2006).

## 13.3  Stereology

### 13.3.1  Description

Stereology is an unbiased and effective tool to obtain quantitative geometric properties such as number, length, area and volume from recorded series of sections of a lower dimension (Vedel Jensen, 1998). It is based on the work of the French mineralogist, Delesse (1848), who applied it to rocks in order to determine the volume of a number of minerals. He showed that the areal fraction occupied by a given mineral on the section of a rock $A_A$ is proportional to the volume fraction of the mineral within the rock volume $V_V$. This notation identifies the frame $A$ or $V$ as the object dimension of interest, and the subscripts as the dimension of the sampling unit. Stereology can typically be used to measure a physical object in $n$ dimensions from random measured objects in fewer dimensions (Baddeley, 1991; Baddeley and Vedel Jensen, 2005). A basic principle of stereology is Cavalieri's principle (Cavalieri, 1635), stating that *"if two solid objects have equal plane sections on all the intersecting planes ($A_h$, $h \in T$), then the objects have equal volume"*. The principle of stereological measurement takes the geometry and probability statistics of an object into consideration.

Two important concepts in stereology are its unbiasedness and its efficient way of taking samples. Unbiasedness implies that, by taking enough samples, the calculated property of the volume gives the population value, whereas sampling concerns the choice of the sampling objects within the object frame. This makes a distinction between classical and modern stereology.

Classical stereology, or model-based stereology, is based on the assumptions that materials are homogeneous in composition. It considers a geometric assumption about the structure of the feature of interest (Baddeley, 1991).

Design-based stereology, which is effective and suited to estimate global and non-homogeneous populations (Baddeley and Vedel Jensen, 2005), is characterized by the fact that no assumptions are made regarding the geometric aspect of the feature of interest, such as its shape, size, and orientation. Design-based sampling procedures are used (Glaser and Glaser, 2000). For example, a test line is selected randomly and the sample is assumed to be arbitrary and fixed. This ensures that each feature of interest in the specimen has an equal probability of being sampled (Baddeley and Vedel Jensen, 2005). Modern stereology enables us to have an unbiased estimate of a higher precision of some geometric quantities (volume, surface area, etc.).

Stereology has been used as a precise, simple and efficient means of quantifying 3D microscopic structures from 2D quantitative sections. Applications of stereology are useful in a broad variety of disciplines, such as in biological, medical, material science, and food science. For instance, in biomedical science, quantitative information about 3D microscopic structures has been based on 2D observations, e.g. to obtain a deeper understanding of the structure and function of the human body and a more objective diagnosis of progressive disease assessment (Roberts *et al.*, 2000).

### 13.3.2 Identifying 2D objects from a remote sensing image

In this study we use stereology within an image mining context to handle the large amount of collected data. We consider 2D objects, of which we estimate the area, and in doing so we use random lines. Although stereology is well described elsewhere (Baddeley, 1991; Baddeley and Vedel-Jensen, 2005), it is less well known in geoinformation science and hence we aim to give a short introduction here.

Stereology extracts structural quantities from measurements made on 2D images: the surface area ($S$) of a 2D object, the length of lines ($L$) and number of points ($N$). We apply the standard notation that points to the objects together with their respective subscripts of estimation to indicate that the measurements are made with respect to this object: $S_V$ is a specific surface area of a volume (i.e., the area per unit volume) and $L_V$ is a specific line length (length per unit volume) of a curve or line structure. This also applies to the length per area $L_A$ and the number of points per area unit, and also to the number of points per linear unit. In this way, the perimeter of an object on an image can be determined from the number of times it meets a straight line placed randomly on an image. Likewise, the length of a curve can be determined from the number of intersection points it makes with a random line.

We now consider an individual remote sensing image and take an object as the region of interest. Sampling is determined by different factors, such as the presence of clouds, day or night at the time of coverage by the satellite, and atmospheric quality, and hence has a random element. In the space-time domain, however, the images are in parallel, as each image is taken at one single moment.

For stereological calculations, an image is first transformed into a binary image, showing the object of interest in black, and the surrounding area in white (Figure 13.1). Next, for $L_L$ estimation, random lines are computed over this transformed image, by determining a random angle within a Python environment. All black pixels on these random lines are counted, resulting in the fraction of each line that falls over the black object in the binary image. This fraction then yields the area $A_A$ of the object in relation to the area of the whole image. The number of random lines can be chosen freely, and a larger number leads to higher precision, but at the expense of increased computation time. In our study we repeated the calculations for several numbers of random lines, e.g. for 10, 100, 1000, 5000 lines, respectively.

## 13.4  Application

This application considers the flooding of the Tonle Sap Great Lake in Cambodia (Fujii et al., 2003). The study area lies on the lower part of the Mekong region of the Cambodian floodplain, following the Mekong river. Flooding due to high water levels of the Mekong River and its tributaries recurs annually. It causes considerable damage to human settlements, agricultural activities and infrastructures of the surrounding area. The Tonle Sap Great Lake covers a relatively small area in the dry season and increases to three to four times this area during the wet season.

We bounded the main study area within a box with the co-ordinates 12° 06′ 25″ N to 13° 55′ 56″ N and 102° 29′ 13″ E to 104° 28′ 51″ E. The country is characterized by five distinct topographic features: the sandstone Dangrek range in the north, forming the border with Thailand, the granite Cardamom Mountains with peaks of over 1500m in the south-west, the Darlac Plateau which rises to over 2700m and in the north-east and the Central Plains between 10 and 30m above sea level, which form 75% of the Cambodian land area.

As we apply remote sensing methods to a flooded area in a densely vegetated area, our object of study is characterized by a reflectance that deviates from that of its neighbourhood. Beginning in May 2001 to January 2002, a series of nine Landsat 7 ETM+, multi-spectral images were used. Each image used in this study was

7667 columns by 6250 rows, with each pixel covering 30m by 30m. The area extent of the lake varies depending on the observation time. Here we consider the images collected on $3^{rd}$ August (rising flood) and on $23^{rd}$ November (peak to falling flood). On the November $23^{rd}$ image we notice that the size of the flooded area is approximately the same as at August $3^{rd}$. The main difference is that it now concerns one large object, with two smaller objects nearby.

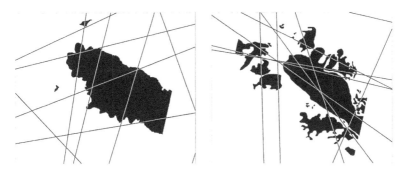

**Figure 13.1.** Representations of the flooding around the Tonle Sap Great Lake in Cambodia from Landsat images taken at $23^{rd}$ of November (left) and $3^{rd}$ of August (right) (Yifru, 2006). Ten random lines are drawn.

The stereological test system applied in this study consists of test lines, a known frame (i.e., the full images) and an incidentally recorded satellite image for information extraction (Figure 13.1). In this particular stereological test system, we equate the flood to our feature of interest $Y_t$ in the spatial domain $R^2$, and we equate the survey area with $X_t$. Apparently, $X_t \subset R^2$ contains a subset $Y_t \subset X_t$.

From Table 13.1 we notice that on November $23^{rd}$ the average size of the lake as estimated by 1000 random lines equals a fraction of 0.1887 of the image, with fractions ranging between 0.1820 and 0.1997. The standard deviation expressed as a fraction is equal to approximately 0.002. Also, the average of the simulations is very close (within the average $\pm 2 \times$ standard deviations) to the histogram based, common method of area estimation, which on that day equals 0.1891 as a fraction of the total image. After converting these fractions into area sizes, we notice that an average size equals 8137km$^2$, corresponding to the histogram based value equal to 8154km$^2$. Precise as it may seem, however, we also notice that the mean error equals 16km$^2$, whereas errors as large as 457km$^2$ occurred. For August $3^{rd}$, similar conclusions were drawn, although the average size of the error expressed in area of land is bigger (60km$^2$), whereas the maximum occurring error is smaller (247km$^2$) on that day than on November $23^{rd}$.

**Table 13.1.** Estimates obtained with 1000 random lines; the error is expressed as the difference between expected and worst estimate of 1000 simulations.

| Date | | Mean | Min. | Max. | Std. Dev. | Observed |
|---|---|---|---|---|---|---|
| 23/11/ 2001 | Fraction | 0.1887 | 0.1820 | 0.1997 | 0.0027 | 0.1891 |
| | Area (km$^2$) | | | 8611 | | 8154 |
| | Error (km$^2$) | 16 | | 457 | | |
| 3/8/ 2001 | Fraction | 0.1943 | 0.1884 | 0.2014 | 0.0024 | 0.1957 |
| | Area (km$^2$) | | | 8687 | | 8440 |
| | Error (km$^2$) | 60 | | 247 | | |

One question that we tried to answer with this study was the number of simulations necessary to do these calculations. Hence we did the calculations also with fewer random lines, leading to the results shown in Table 13.2. We notice that the standard deviations increase (from 0.0027 to 0.0565), as well as the interquartile range, when the number of simulations drops from 5000 down to 10. Changes of the mean area fractions size are less clear: on November 23rd we notice a decrease in size from 0.1887 to 0.1802, whereas on August 3rd an increase in size from 0.1943 to 0.2014 occurs with the decreasing number of lines.

**Table 13.2.** Mean, standard deviations (std) and interquartile ranges (IQR) for estimated fractions of inundations for different numbers of simulations (5000, 1000, 100 and 10 lines, respectively).

| Date | Observed | | 5000 | 1000 | 100 | 10 |
|---|---|---|---|---|---|---|
| 23/11/ 2001 | 0.1891 | Mean | 0.1887 | 0.1899 | 0.1871 | 0.1802 |
| | | Std | 0.0027 | 0.0062 | 0.0181 | 0.0565 |
| | | IQR | 0.0033 | 0.0091 | 0.0244 | 0.0772 |
| 3/8/ 2001 | 0.1957 | mean | 0.1943 | 0.1943 | 0.1953 | 0.2014 |
| | | std | 0.0024 | 0.0050 | 0.0147 | 0.0541 |
| | | IQR | 0.0032 | 0.0060 | 0.0186 | 0.0792 |

## 13.5  Discussion

Flooding can be characterized using image mining. It begins at a particular moment in time, it may be poorly visible at several moments, because of (partial) cloud cover, it increases in size, it may split, several objects may merge and, after the river withdraws, the flood ends and the object reduces to its original size. Moreover, the boundaries between flooded and non-flooded land are relatively simple to draw, although (partial) cloud cover may prohibit its precise and detailed observation. In this study, the accuracy of stereologically estimated 2D flooded areas by passing different sets of random lines (10, 100, 100, and 5000) depends upon the number of lines drawn to quantify the flooded area, and the lines passing through the image should be selected at random. Uncertainty arises from different sources. One source of uncertainty is knowledge uncertainty, referring mainly to the input data and the assumptions. A second source of uncertainty is the dimension of pixels in the image.

Estimators for the fraction of lines within the flooded area are the numbers of pixels. In principle, we could replace this by distributing points at random over the area, and counting the number of times that these points fall within the area. At this stage we have restricted ourselves to the case of random lines, however, which are already showing various interesting results.

Spatial data quality affects GIS based decision-making activity (Shi *et al.*, 2002). When using data as an input in numerical models, data precision should be taken care of, in particular if it propagates towards the decision. Spatial data quality refers to various aspects of data quality as can be identified for geographical objects. We restrict ourselves to the two main components: namely positional accuracy and attribute accuracy. Apparently the two are closely related. Use of stereology allows us to estimate both the size and a measure of accuracy. At this stage we can recommend for areas similarly identifiable as a lake to take 1000 random lines. The additional value of taking more than this number seems to be limited, as the area sizes are similar and standard errors are low.

In this study we have focused on crisp objects. There seems to be little doubt that stereology can be extended as well to uncertain objects. Then, $\alpha$-shapes, membership functions or random set based properties should be related to the random lines now considered. We leave this for further research.

## 13.6 Conclusion

We conclude that stereology is applicable to remote sensing images on flooding. This is partially thanks to the crisp nature of flooded objects. A limited number of random lines appears to be sufficient for defining the area of a flooded object. The derived area is apparently close to the area in reality although, in terms of square kilometres, differences are substantial. In a subsequent paper, we will investigate the issue further, i.e. making a comparison with a larger set of images and extending the procedures in space and time.

## References

Baddeley, A. (1991), "Stereology". *In: Spatial Statistics and Digital Image Analysis*, chapter 10. National Research Council USA, Washington DC, pp. 181-216.

Baddeley, A. and E.B.Vedel Jensen. (2005), *Stereology for Statisticians*. Chapman and Hall/CRC 395, London, U.K., 395p.

Cavalieri, B. (1635), *Geometria indivisibilibus continuorum. Bononi: Typis Clementis Ferronij*. Reprinted 1966 as *Geometria Degli Indivisibili*. Unione Tipografico-Editrice Torinese, Torino, Italy, 872p.

Delesse, M.A. (1848), "Procédé mécanique pour déterminer la composition des roches". *Annales des Mines*, Vol.13: 379-388. Quatrième série.

Dilo, A., R. De By, and A. Stein. (2007), "A system of types and operators for handling vague spatial objects". *International Journal of Geographical Information Science*, Vol. 21(4): 397-426.

Frank, A.U. (2008), "Analysis of dependence of decision quality on data quality". *Journal of Geographical Systems*, Vol. 10(1): 71-88.

Fujii, H., H. Garsdal, P. Ward, M. Ishii, K. Morishita, and T. Boivin. (2003), "Hydrological roles of the Cambodian floodplain of the Mekong River". *International Journal of River Basin Management*, Vol. 1(3): 1-14.

Glaser, J. and E.M. Glaser. (2000), "Stereology, morphometry, and mapping: the whole is greater than the sum of its parts". *Journal of Chemical Neuroanatomy*, Vol. 20: 115-126.

Lucieer, A., A. Stein, and P. Fisher. (2005), "Multivariate texture segmentation of high-resolution remotely sensed imagery for identification of fuzzy objects". *International Journal of Remote Sensing*, Vol. 26: 2917-2936.

Rajasekar, U., A. Stein, and W. Bijker. (2006), "Image mining for modeling of forest fires from Meteosat images". *IEEE Transactions on Geoscience and Remote Sensing*, Vol. 45(1): 246-253.

Richards, J.A. and X. Jia. (1998), *Remote sensing digital image analysis - third edition*. Springer, Berlin, 363p.

Roberts, N., M.J. Puddephat, and V. McNulty. (2000), "The benefit of stereology for quantitative radiology". *British Journal of Radiology*, Vol. 73: 679-697.

Shi, W., P.A. Fisher, and M. Goodchild. (2002), *Spatial Data Quality*. Taylor & Francis, London, U.K., 313p.

Shi, W. (2009), *Principles of Modeling Uncertainties in Spatial Data and Spatial Analyses*, CRC Press, Boca Raton. 432p.

Silva, M.P.S., G. Câmara, M.I.S. Escada, and R.C.D. de Souza. (2008) "Remote sensing image mining: detecting agents of land use change in tropical forest areas". *International Journal of Remote Sensing*, Vol. 29: 4803-4822.

Stein, A. (2008), "Modern developments in image mining". *Science in China Series E: Technological Sciences*, Vol. 51, Supplement 1: 13-25.

Stein, A., W. Shi, and W. Bijker. (2008), *Quality aspects in spatial data mining*. CRC Press, Boca Raton, 364p.

Stoyan, D. and H. Stoyan. (1994), *Fractals, random shapes and pointfields*. John Wiley & Sons, Chichester, 386p.

Van de Vlag, D. and A. Stein. (2006), "Uncertainty propagation in hierarchical classification using fuzzy decision trees". *IEEE Transactions on Geoscience and Remote Sensing*, Vol. 45(1): 237-245.

Vedel Jensen, E.B. (1998), *Local Stereology*. World Scientific Publishing, Singapore. 272p.

Yifru, M.Z. (2006), *Stereology for Data Mining*. Unpublished MSc thesis, ITC International Institute for Geoinformation Science and Earth Observation, Enschede, The Netherlands. 57p.

Zadeh, L. (1965), "Fuzzy sets", *Information and Control*, Vol. 8: 338-353.

# 14

# Assessing Digital Surface Models by Verifying Shadows: A Sensor Network Approach

*Cláudio Carneiro, Mina Karzand, Yue M. Lu, François Golay & Martin Vetterli*

## Abstract

*This paper proposes the use of wireless sensor networks to assess the accuracy and application of Digital Surface Models (DSM) for the study of shadowing and solar radiation over the built environment. Using the ability of sensor network data to provide information about solar radiation and predicting the exact time of the day that the Sun starts radiating a sensor, a comparative study and statistical analysis can be undertaken in order to evaluate the accuracy of the DSM for shadowing and radiation studies using image processing techniques. Two DSMs of the EPFL campus with different cell resolutions (1 meter and 0.5 meters), considering only information about ground, buildings with vertical walls and trees, are constructed step by step and employed. Three DSMs of the same campus with a cell resolution of 1 meter derived from raw LIDAR data and common interpolation techniques – Triangulated Irregular Network (TIN), kriging, Inverse Distance Weighting (IDW) – are also used for comparison.*

## 14.1 Introduction

Today's availability of 3D information concerning cities enables the study of urban frameworks with new and more sophisticated approaches. In fact, even though airborne Light Detection and Ranging (LIDAR) data permits the derivation of valuable and accurate information about the physical arrangement of cities, few applications have been implemented in order to process this data for environmental analysis, such as shadowing and solar radiation, and hence, understanding the performance of the urban form. For instance, the growing importance given to the quantification of energy-based indicators at the scale of the city strongly suggests the incorporation of 3D geography in order to provide useful applications for urban planning (Osaragi and Otani, 2007).

The analysis of shadowing and solar radiation in architecture and urban planning is a well-studied problem (e.g., Morello and Ratti, 2009). In fact, tools that can accurately calculate the radiation performance of buildings already exist, such as the RADIANCE lighting simulation model (Ward, 1994). Nevertheless, these tools are very useful at the micro-scale of architecture (environmental performance software) or at the macro-scale of landscape and regional geography (GIS tools), but the focus on urban districts and tools for automatically calculating irradiation on a whole district are mostly lacking.

Airborne Laser Scanning (ALS), also called LIDAR, is a modern, fast and high-resolution technology that integrates sensors in order to obtain very accurate 3D coordinates – XYZ data points located on the surface of the earth – such as ground points, buildings and trees. In order to establish the position of the sensor each time a point is measured, a Global Positioning System (GPS) is used. For finding out the attitude of the sensor, an Inertial Navigation System (INS) is adopted by using a narrow laser beam to determine the range between the sensor and the target points.

Thus, using the capability of raw point clouds data derived from LIDAR technology and vector building footprints stored in a GIS, an accurate 2.5D urban surface model, also called a Digital Surface Model (DSM), can be constructed and further applied to the calculation of shadowing and radiation on urban fabric at the neighbour/district scales.

In this paper we intend to validate the accuracy of the five gridded DSM of the EPFL campus, Lausanne, Switzerland, generated by different interpolation methods and techniques. The shadowing (i.e., solar radiation discontinuities) predicted by these DSMs are compared to the real solar radiation measurements obtained by a wireless sensor network deployed on the same campus.

Finally, we compute the average error between the transition time derived from the solar radiation measurements and that from shadowiness analysis using a DSM, validating the usefulness of DSMs for this type of environmental study.

## 14.2  Related Work

Previous literature on the interpolation of LIDAR point clouds is not rare. The advantages and disadvantages of numerous interpolation methods, such as triangle-based linear interpolation, nearest neighbor interpolation and kriging interpolation were presented by Zinger *et al.* (2002). Control techniques that analyze the precision, accuracy and reliability of digital terrain models (DTM) and DSM can be carried out for applications where a high level of accuracy is demanded (Menezes *et al.*, 2005; Gonçalves, 2006; Karel *et al.*, 2006).

The effects of LIDAR data density on the accuracy of digital elevation models and the extent to which a set of LIDAR data can be reduced yet still maintain adequate accuracy for DEM generation were studied by Liu and Zhang (2008). Aspects also studied include the source and nature of errors in digital elevation models, and in the derivatives of such models (Fisher and Tate, 2006); and the recent advances of airborne LIDAR systems and the use of LIDAR data for DEM generation, with special focus on LIDAR data filters, interpolation methods, DEM resolution, and LIDAR data reduction (Liu, 2008) As related research, a quality improvement of laser altimetry of digital elevation models using ground control points in a 1D strip adjustment was proposed by Oude Elberink et al. (2003).

A method to interpolate and construct a DSM (incorporating the geographical relief), based on LIDAR and GIS buildings data, has already been proposed by Osaragi and Otani (2007). Also, two different studies concerning the analysis of solar potential of roofs using DSM have been recently presented (Kassner *et al.*, 2008; Beseničara *et al.*, 2008). As related research, there are some semi-automatic methods available to create 3D GIS data from LIDAR data, such as the Virtual London at the University College London (Steed *et al.*, 1999) and the MapCube at Tokyo CadCenter Corporation (Takase *et al.*, 2003).

Furthermore, a group of researchers at the Martin Centre, University of Cambridge can be considered pioneers in the use of DSM to extract environmental indicators, as far as literature about the application of raster images in urban studies is concerned (Ratti, 2001; Ratti and Richens, 2004; Ratti *et al.*, 2005).

Today's growing accessibility of 3D information from user-generated content and remote sensing surveys, makes this technique extremely useful for a common understanding of the performance of our cities. However, there is a lack of information concerning the accuracy of these models for this type of environmental analysis. Thus, in this work we aim to assess the accuracy of shadowing predictions made using these DSM through the comparison with real solar radiation measurements directly derived from sensor network data.

## 14.3 Construction and Gridding Interpolation of Digital Surface Models (DSM)

### 14.3.1 Presentation

The goal of gridding interpolation techniques is to generate, through spatial interpolation, a rectangular array of Z values derived from irregularly spaced XYZ data points. Many spatial interpolation methods are available and can be classified, including:

- **Global**: each interpolated value, defined as a cell node of the gridded DSM, is influenced by all the data points, in this case raw LIDAR data in the form of XYZ point clouds.
- **Local**: each interpolated value is only influenced by the values at predefined nearby points of the XYZ point clouds.
- **Exact**: creates a surface that passes through all of the XYZ point clouds.
- **Approximate**: produces a surface that follows only an overall tendency in the XYZ point clouds, in which there exists a few degrees of error.
- **Stochastic**: incorporates geo-statistic theory in order to produce surfaces with particular levels of error.

In our testing areas at the EPFL campus, LIDAR points were obtained with a density of one point per square meter and only one LIDAR pulse. As presented by Behan (2000), the most accurate surfaces are achieved using a grid with a sampling size that matches as much as possible the LIDAR point density during the acquisition phase. Hence, four of the five gridding interpolation techniques here presented have a sampling size of one meter.

### 14.3.2 Gridding interpolation techniques

For large sets of LIDAR point clouds the grid computation of some global interpolation methods is too slow. However, these interpolation methods can be transformed into local by limiting the interpolation area to a neighbor area. Thus, in order to implement this project we gridded the following DSM with a sampling size of one meter:

- Three DSMs derived from common local interpolation methods: inverse distance weighting (global interpolation), kriging (stochastic interpolation) and triangulation with linear interpolation (exact interpolation).
- One DSM, called 2.5D Urban Surface Model (2.5D USM): the interpolation technique is applied in different steps, as presented in Section 14.3.2.

A fifth DSM (second 2.5D USM), with a sampling size of 0.5 meters, was also used for comparison.

Finally, all interpolated DSMs were slightly smoothed by applying a 3 by 3 low-pass filter.

### 14.3.3  Inverse Distance Weighting

Quite often the inverse distance weighting technique is used for interpolation of irregularly spaced points. In this method, the LIDAR point clouds are weighted during interpolation in order to decrease the influence of one point relative to another with distance from the grid node under analysis (Shepard, 1968). The main concept inherent to this technique is that nearby points have similar height values, while the heights at distant points are classified as being independent. Moreover, a weighting power that controls how the weighting factor drops off as distance from a grid node increases is usually assigned to data.

The generation of 'bull's-eyes' surrounding the position of observations within the gridded area occurs after applying inverse distance weighting interpolation. Thus, a smoothing parameter can be applied during the interpolation process in order to reduce the bull's-eye output by smoothing the interpolated grid.

### 14.3.3.1  Kriging

Kriging is a geo-statistical interpolation technique that allows us to estimate the heights at the grid nodes as a weighted average of the measured heights at the reference points (LIDAR point clouds), usually in two steps: weight determination and the estimation of the height values using a weighted average. A procedure called variogram modeling, which describes the spatial variability between the height values of the reference points, is used for the determination of weights (Cressie, 1993).

### 14.3.3.2  Triangulation with linear interpolation

The applied algorithm creates a triangulated irregular network (TIN) structure from the LIDAR points using a Delaunay Triangulation routine, which maximizes the minimum angle of all the angles of the triangles in the triangulation undertaken. The original points are connected in such a way that no triangle edges are intersected by other triangles. A sequential search allows the set up of a triangle in which each grid node is enclosed. The gradients of the chosen triangle enable the interpolation of a value for the grid node (Franklin, 1973).

### 14.3.3.3  Construction and interpolation of a 2.5D Urban Surface Model

Two data sources are required for the interpolation of the 2.5D Urban Surface Model (2.5D USM) of the EPFL campus: raw LIDAR data and 2D digital vector maps of buildings footprints stored in a GIS.

First, we interpolate a digital terrain model (DTM) by classifying the LIDAR points. Using GIS software, LIDAR points confined within building polygons and in the two meters buffer generated from building polygons are eliminated. In addition, in a neighborhood of two meters, LIDAR points whose elevation value varies significantly from surrounding points are considered to be points indicating features such as aerial points (e.g., if the laser beam touches a bird) and vehicles, and thus are removed. After eliminating the points based on the features described above, a DTM can be interpolated from the ground points only. Due to its generalized use by the scientific community for DTM interpolation, the triangular interpolation (construction of a TIN) was chosen.

Second, using only the LIDAR points classified as being contained within vector buildings' footprints, a triangular interpolation (construction of a TIN) is independently applied to each of these buildings' roofs. It is important to note that, along the edges of the roofs, there exist some laser points touching walls and not roofs, which may influence the TIN generation of each roof. Thus, wall points need to be detected. A building point is classified as wall point if there are much higher points and none or very few points that have its approximate height within a neighborhood of two meters. Finally, construction of a 2.5D surface model of building roofs is applied.

According to the algorithm initially presented by Axelsson (1999), LIDAR points considered to be trees higher than five meters are classified. Hence, construction of a 2.5D surface model of trees is also applied.

For each grid cell considered to be a building or tree, its height (also defined as nDSM of buildings and trees) is taken to be the value of the difference between the terrain elevation (calculated in the interpolated DTM) and the building and tree elevations.

Lastly, each building and tree is added to the DTM, using the height found previously for each cell contained within, as described in the last paragraph. The final result allows the construction of a 2.5D USM: DTM + nDSM of building roofs + nDSM of trees.

Finally, in order to complete the image enhancement of the model, we have to refine the facades of buildings that are slightly sloped because of interpolation. Thus, each building's contour pixel was deleted and then expanded again, in order to assign more constant values to roof edges.

A blackbox describing the implemented methodology for the construction of the 2.5D USM is shown in Figure 14.1. Data sources and parameters needed to generate the 2.5D USM are shown in Figure 14.2.

An example of a 2.5D USM of the EPFL campus, at the city of Lausanne, Switzerland, is shown on the right-hand side picture of Figure 14.3.

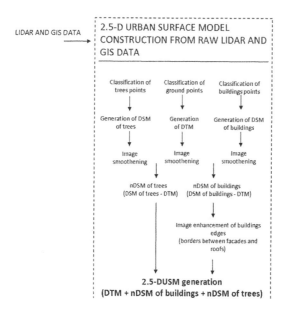

**Figure 14.1.** Blackbox describing the implemented methodology for the construction of the 2.5D USM.

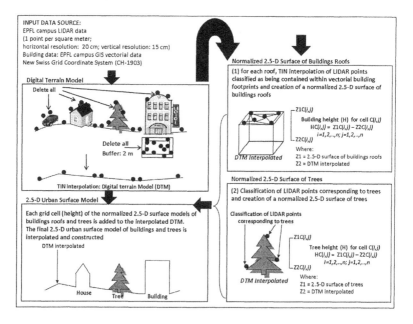

**Figure 14.2.** Technique applied for the construction of a 2.5D urban surface model (2.5D USM) with information of ground, buildings with vertical walls and trees.

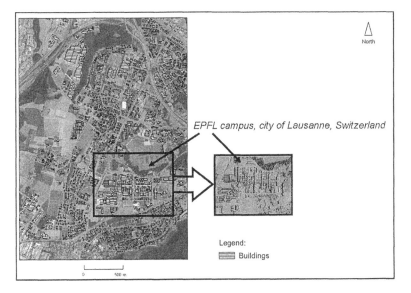

**Figure 14.3.** Picture on the left-hand side: GIS building footprints and aerial pictures of the district of Chavannes and EPFL campus, city of Lausanne, Switzerland; picture on the right-hand side: 2.5D urban surface model (2.5D USM) of the EPFL campus, city of Lausanne, Switzerland; black rectangle: case study (pilot zone), within the EPFL campus, city of Lausanne, Switzerland.

## 14.4 SensorScope Project

As presented by Barrenetxea *et al.* (2008), SensorScope is a joint project between network, signal processing, and environmental researchers that aims to provide a cheap and out-of-the-box environmental monitoring system based on a wireless sensor network. It has been successfully used in a number of deployments to gather hundreds of megabytes of environmental data. The geographical position of the wireless sensor network of the EPFL campus was originally defined in WGS84 coordinates (GPS measurements) and later transformed into New Swiss Grid (CH-1903 datum) coordinates. The location and labeling (identification) of each sensor of the EPFL campus wireless sensor network is shown in Figure 14.4.

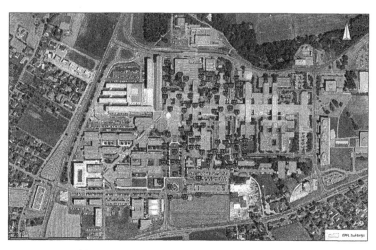

**Figure 14.4.** Position and identification of the wireless sensors network at the EPFL campus, city of Lausanne, Switzerland.

## 14.5 Presentation of the Accuracy Validation Methodology

In order to validate the accuracy of the DSM presented, the predictions based on the map are compared with real measurements of solar radiation measured by sensors. The solar radiation parameter depends greatly on the amount of existing clouds during the day: on cloudy or rainy days the density of clouds in front of the sun determines how much solar radiation is received by the sensor. Thus, as a result of the disordered nature of the water drops in clouds, the measured solar radiation is a very chaotic signal. Conversely, on sunny days, measured solar radiation is a piece-wise smooth signal which increases from morning to noon and decreases from noon to evening. The only irregularity in this signal is some jumps and drops in the level of the measured solar radiation. By jumps and drops, we mean there is a sudden increase or decrease in the level of signal over a very short time interval. The reason of existence of these is shadowing phenomena, i.e., whenever a sensor gets out of shadow, the solar radiation measured by it increases a lot in a short amount of time, and whenever it goes into the shadow, the measured solar radiation drops suddenly. The main idea of the project was to use the DSM to predict the exact times of changing of shadowing for each sensor position (XYZ coordinates) on each sunny day and compare this prediction to the jumps and drops of the sensor measurements, such as presented in Figure 14.5.

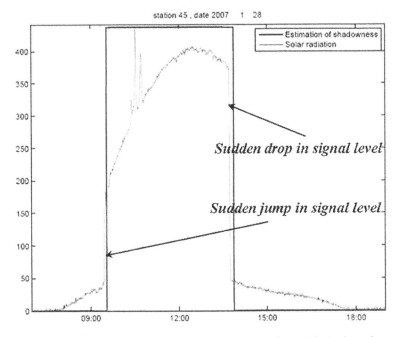

**Figure 14.5.** Comparison, on a sunny day, between the estimated shadowiness for a sensor position (XYZ coordinates) using a DSM and the solar radiation information from the same sensor.

As shown in Figure 14.5, there is no perfect match between the two independent parameters: estimation of shadowiness for a sensor position (XYZ coordinates) using DSM and solar radiation measured by sensors. Due to the fact that the prediction of shadowiness uses the DSM as its input data, and as we may be somewhat confident about the times of jumps and drops read from sensor measurements, the error will most likely be caused by the existing errors of the DSM. Thus, this error might be considered as a good measure of the quality of the DSM. A back tracing algorithm was used to estimate the shadow-state of each sensor at each time instant. Based on the day of the year and the latitude and longitude of the city of Lausanne, the algorithm calculates the exact direction of the Sun for each sensor at each moment and back-traces the ray of light from the sensor toward the Sun. Then, the height of this ray of light is compared to the curve derived from the DSM, which shows the elevation along the line in the same direction, as shown in Figure 14.6.

If there is any intersection where the ray of light is below the elevation according to the DSM, the point of that intersection will cast shadow on the corresponding sensor at that moment.

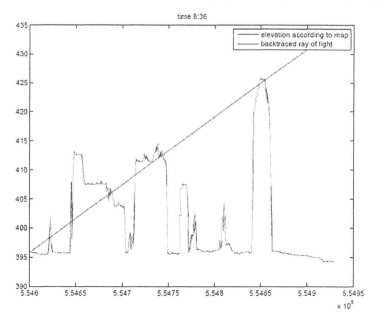

**Figure 14.6.** Comparison between the height of the ray of light, which is back-traced towards the Sun, and the elevation values of the DSM.

An illustrative representation of a theoretical shadow-map of EPFL campus at one time instant is presented in Figure 14.7. The direction of light coming from the sun is also shown in this figure.

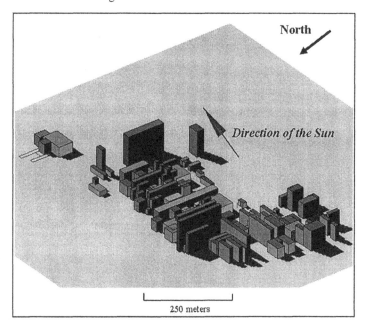

**Figure 14.7.** For one time instant and the direction of sun, illustrative presentation of the theoretical shadow-map in the EPFL campus.

## 14.6 Analysis of Results

The prediction of shadowiness is undertaken according to different DSMs and the measurements of solar radiation for each sensor on eight sunny days distributed throughout the year. Two parameters are introduced as a measure of accuracy (deviation and number of jumps and drops) between the application of DSM for shadowiness analysis for a sensor position (XYZ coordinates) using image processing techniques and the solar radiation information of sensors:

- **Number of jumps and drops** that could be predicted approximately for a threshold of 30 minutes. This threshold was defined in order to exclude all cases of sensors presenting an unstable behavior concerning solar radiation measurements. For most of the cases these sensors are at short distances from trees, where the shadowiness factor caused by the natural movement of tree leaves is not negligible. Moreover, the classification and interpolation of trees is a complex issue on DSM interpolation. In this case mainly because the density of one point per square meter of LIDAR points available is too low. For this reason, the estimation of shadowiness caused by trees using DSM image processing techniques is also highly influenced by the quality of the interpolated DSM. In Figure 14.8 an example of a sensor with a fuzzy performance that could not be used for this classification is shown.

- **Average deviation error (in minutes)** for those jumps that could be predicted for a threshold of 30 minutes. The error is defined as the difference of time between jumps in solar radiation and boundaries of shadowiness, which are predicted according to the DSM.

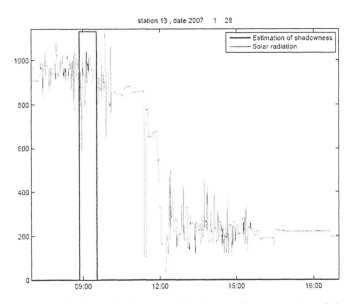

**Figure 14.8.** Example of a sensor with a fuzzy performance: solar radiation line.

By using the timing errors of prediction of shadow and some geometrical assumptions on the environment, a spatial error parameter could also be defined. The imperfect matching of the prediction and sensor readings are because of error in interpolated value for the elevation of dome points. By having the angle of sun in the predicted time and in the real time of the shadow transition and the vertical distance of the sensor and the obstruction point, the error in height of the obstruction point could be estimated. Corresponding to Figure 14.9, assume that the position of the sensor is at point $O$. Based on the information in the DSM, point $A$ is assumed to cast a shadow on the sensor at time $t_1$. So, the angle of sun at this time is $\Phi$, and the ray of light is assumed to be the line $AO$. However, the measurements show that the shadow transition happens at time instant $t_2$, in which the sun has the angle of $\psi$ for the point $O$. Thus, the real ray of light that causes the transition in shadow state of point $O$ is $BO$. The error in estimating the shadow transition is because of the error in defining the elevation, which is $AB$. Using simple geometry, we can define the spatial error as:

$$\text{Spatial error} = |AB| = (\tan(\psi)-\tan(\Phi)) \times d$$

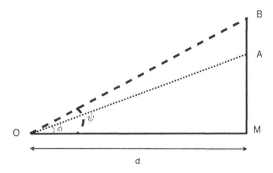

**Figure 14.9.** The model used for estimating the spatial error on the vertical direction of the DSM. The sensor is at point $O$. The model predicts the transition to happen at time $t_1$ when the sun has horizontal angle $\Phi$, but the real transition happens at time $t_2$ corresponding to angle $\psi$.

Thus, for each transition in the shadow state, a temporal error and a spatial vertical error could be defined. The simulation results of the average deviation error (in minutes), the number of jumps and drops, and the average vertical spatial error (in centimeters) for each of the five DSM under analysis are presented in Table 14.1.

**Table 14.1.** Average Deviation Error: AvDE (minutes), Number of Jumps and Drops: NJD, and Average Vertical Spatial Error: AvSE (centimeters) for each DSM under assessment.

| DSM | AvDE (minutes) | NJD (threshold: 30 minutes) | AvSE (cm) |
|---|---|---|---|
| 2.5D USM (1m of resolution) | 4.57 | 85 | 26 |
| 2.5D USM (0.5m of resolution) | 5.14 | 85 | 16 |
| IDW (1m of resolution) | 6.17 | 76 | 22 |
| Kriging (1m of resolution) | 5.19 | 77 | 19 |
| TIN (1m of resolution) | 6.19 | 85 | 19 |

From the analysis of Table 14.1 we can observe that the average deviation error of the 2.5D USM with one meter of resolution presents slightly better results than all the other DSM. In this case, according to the construction method applied and contrary to the other interpolation methods used (IDW, Kriging and TIN), the edges of roofs are more accurately defined. Hence, this will affect the propagation of shadowiness over the built environment using the DSM (2.5D USM) and will also implicitly improve its final accuracy. The 2.5D USM with a sampling size of 0.5 meters presents slightly worse results than the 2.5D USM with a sampling size of one meter. Once again, this can be caused by the fact that the density of LIDAR points used for this classification is equal to one point per square meter, and according to Behan (2000) the most accurate surfaces are created using a grid with a sampling size that relates as close as possible to the LIDAR point density during the acquisition phase.

It is noteworthy that, as the mapping between the vertical spatial error and the temporal error can differ markedly from one sensor to the other, there exists no simple relationship between the average vertical error and the average temporal error reported in Table 14.1. The average vertical spatial error is strongly influenced by the distance parameter. Thus, the 2.5D USM with a higher resolution, here constructed with a sampling size of 0.5 meters, presents slightly better results than all the other DSMs under analysis, which have a coarser sampling size of one meter.

## 14.7 Conclusions

Using solar radiation measurements taken by a wireless sensor network, we present an algorithm for evaluating the quality and usefulness of five different DSMs in shadowiness and solar radiation studies on urban areas.

The results of this work show that there is no great difference among the four interpolation methods using a grid with a sampling size of one meter. The 2.5D USM presented here is the most robust method, and for this reason its use should be generalized for this kind of environmental applications at neighbor/district city scales.

It was found that existing vector digital maps (GIS data) can be used if available and updated, but outlines of buildings from this source of information should always be handled with special care. In fact, the 2D outlines of building footprints do not necessarily represent the outline of the building roof. Modifications between GIS data and laser data can have numerous impacts, which cannot be automatically recognized. Proposals using vector digital maps as input for 2.5D urban surface model interpolation and construction should be attentive of the fact that, in some cases, map information might not give the correct hints about 3D building shapes.

Finally, some improvements concerning the technical implementation of the presented methodology are needed in future work, especially:

- The development of an algorithm for the automatic selection of solar radiation information for sensors presenting a smooth signal (sunny days).

- The use of modern LIDAR datasets with higher density of points per square meter and different signal pulses will achieve better results. Thus, it will be possible to improve the classification and reconstruction of trees (using all pulses) and buildings (using first and last pulses) for the 2.5D USM here presented, and also the reconstruction of DSM with denser and more accurate sampling size, such as 0.5 meters.

## References

Axelsson, P. (1999), "Processing of laser scanner data - algorithms and applications". ISPRS Journal of Photogrammetry and Remote Sensing, Vol. 54: 138-147.

Barrenetxea, G., F. Ingelrest, Y.M. Lu and M. Vetterli. (2008), "Assessing the challenges of environmental signal processing through the SensorScope project", *In: Proceedings of the 33rd IEEE International Conference on Acoustics, Speech, and Signal Processing (ICASSP 2008),* Las Vegas, USA, pp. 5149-5152.

Behan, A. (2000), "On the matching accuracy of rasterised scanning laser altimeter data". *In: International Archives of Photogrammetry, Remote Sensing* and Spatial Information Sciences, Amsterdam, The Netherlands, Vol. 33(B2): 75-82.

Beseničara, J., B. Trstenjak and D. Setnikac. (2008), "Application of Geomatics in Photovoltaics". *In: International Archives of Photogrammetry, Remote Sensing* and Spatial Information Sciences, Beijing, China, Vol. 37(B4): 53-56.

Cressie, N. (1993), Statistics for Spatial Data. John Wiley & Sons, New York, USA, 900p.

Oude Elberink, S., G. Brand and R. Brügelmann. (2003), "Quality improvement of laser altimetry DEM's". *In: Proceedings of the ISPRS working group III/3 workshop "3-D reconstruction from airborne laserscanner and InSAR data",* Dresden, Germany, 34(3/W13).

Fisher, P. and N.J. Tate. (2006), "Causes and consequences of error in digital elevation models". Progress in Physical Geography, Vol. 30(4): 467-489.

Franklin, W.R. (1973), *Triangulated irregular network program.* Downloadable from
ftp://ftp.cs.rpi.edu/pub/franklin/tin73.tar.gz.

Gonçalves, G. (2006), "Analysis of interpolation errors in urban digital surface models created from LIDAR data". *In:* M. Caetano and M. Painho (eds.) Proceedings of the 7th International Symposium on Spatial Accuracy Assessment in Resources and Environment Sciences. Lisbon, Portugal, pp. 160-168.

Karel, W., N. Pfeifer and C. Briese. (2006), "DTM quality assessment". *In:* International Archives of Photogrammetry, Remote Sensing and Spatial Information Sciences, Vol. 36(2): 7-12.

Kassner, R., W. Koppe, T. Schüttenberg and G. Bareth. (2008), "Analysis of the Solar Potential of Roofs by Using Official LIDAR data". *In:* International Archives of the Photogrammetry, Remote Sensing and Spatial Information Sciences, Vol. 37(B4): 399-404.

Liu, X. (2008), "Airborne LIDAR for DEM generation: some critical issues". Progress in Physical Geography, Vol. 32(1): 31-49.

Liu, X. and Z. Zhang. (2008), "LIDAR data reduction for efficient and high quality DEM generation". *In:* The International Archives of the Photogrammetry, Remote Sensing and Spatial Information Sciences, Vol. 37(B3b): 173-178.

Menezes, A.S., F.R. Chasco, B. Garcia, J. Cabrejas and M. González-Audícana. (2005), "Quality control in digital terrain models". Journal of Surveying Engineering, Vol. 131: 118-124.

Morello, E. and C. Ratti. (2009), "SunScapes: 'solar envelopes' and the analysis of urban DEMs, Computers", *Environment and Urban Systems*, Vol. 33(1): 26-34.

Osaragi T. and I. Otani. (2007), "Effects of ground surface relief in 3D spatial analysis on residential environment". *In:* S.I. Fabrikant and M. Wachowicz (eds.), The European Information Society: Lecture notes in Geoinformation and Cartography, Springer, Berlin, Germany, pp. 171-186.

Ratti, C. (2001), Urban analysis for environmental prediction, Ph.D thesis, University of Cambridge, Cambridge, UK, 313p.

Ratti, C. and P. Richens. (2004), "Raster analysis of urban form". Environment and Planning B: Planning and Design, Vol. 31(2): 297-309.

Ratti, C., N. Baker and K. Steemers. (2005), "Energy consumption and urban texture". Energy and Buildings, Vol. 37(7): 762-776.

Shepard, D. (1968), "A two-dimensional interpolation function for irregularly-spaced data". *In*: R.B. Blue, Sr. and A.M. Rosenberg (eds.) Proceedings of the 1968 ACM National Conference, ACM Press, New York, USA, pp. 517-524.

Steed, A., E. Frecon, D. Pemberton and G. Smith. (1999), "The London Travel Demonstrator", *In:* M. Slater, Y. Kitamura, A. Tal, A. Amditis and Y. Chrysanthou (eds.) *Proceedings of the ACM Symposium on Virtual Reality Software and Technology*, ACM Press, New York, USA, pp. 50-57.

Takase, Y., N. Sho, A. Sone, and K. Shimiya. (2003), "Automatic generation of 3D city models and related applications". *In: International Archives of the Photogrammetry, Remote Sensing and Spatial Information Sciences*, Vol. 34-5/W10.

Ward, G.J. (1994), "The Radiance Lighting Simulation and Rendering System" *In:* D. Schweitzer, A. Glassner and M. Keeler (eds.). *Proceedings of the 21st annual conference on Computer graphics and interactive techniques*, ACM Press, New York, USA, pp. 459-472.

Zinger, S., M. Nikolova, M. Roux and H. Maître. (2002), "3-D resampling for airborne laser data of urban areas". *In: International Archives of the Photogrammetry, Remote Sensing and Spatial Information Sciences*, Vol. 34(3B): 55-61.

# 15

# Evaluation of Remote Sensing Image Classifiers with Uncertainty Measures

*Luísa M.S. Gonçalves , Cidália C. Fonte, Eduardo N.B.S. Júlio & Mario Caetano*

## Abstract

*Several methods exist to classify remote sensing images. Their appropriateness for a particular objective may be analysed by evaluating the classification accuracy with indices derived from the traditional confusion matrix, considering a sample of points within the regions used as testing sites. However, this analysis requires the existence of reference data corresponding to the ground truth for the used sample and its application is limited to a sample of points. This paper intends to show that, for classifiers from which uncertainty information may be obtained, the evaluation of the classifier performance can also be made with uncertainty indices. Two supervised classifiers are used, a Bayesian probabilistic classifier and a fuzzy classifier. Their performance is evaluated by using accuracy and uncertainty information and it is shown that similar conclusions may be obtained with both. Therefore, uncertainty indices may be used, along with the possibility or probability distributions, as indicators of the classifiers performance, and may turn out to be very useful for operational applications.*

## 15.1  Introduction

The classification of remote sensing images may be performed with several methods, which have different characteristics and are based on different assumptions. The supervised classifiers perform the classification considering the spectral responses of the image for a sample of points chosen in regions considered representative of the several classes. This information is then used to classify the whole image. To choose the appropriate classifier for each data set and nomenclature, it is necessary to determine which classifiers are the most adequate. Traditionally this is evaluated considering an additional sample of points within the regions representative of the classes and evaluating the accuracy of the classification for those points, constructing an error matrix and computing accuracy indices for the classes, such as the User Accuracy (UA) and the Producer Accuracy (PA), and global indices, such as the Global Accuracy and the KHAT Coefficient. However, this approach has some drawbacks: (1) it requires the identification of the ground truth for the additional sample of points, which is always a time consuming operation (Gopal and Woodcock, 1994; Gonçalves *et al.*, in press); (2) it is influenced by the sampling methods and sample size (Stephman and Czaplewski, 2003); (3) it is influenced by a certain degree of subjectivity of the reference data (Plantier and Caetano, 2007); (4) it is only appropriate for traditional methods of classification, where it is assumed that pixels at the reference locations can be assigned to single classes (Lewis and Brown, 2001); and (5) it cannot provide spatial information of where thematic errors occur (Brown *et al.*, 2009).

In remote sensing, classifiers that provide additional information about the difficulty in identifying the best class for each pixel have been developed, such as the posterior probabilities of the maximum likelihood classifier (e.g. Foody *et al.*, 1992; Bastin, 1997; Ibrahim *et al.*, 2005), the grades of membership of fuzzy classifiers (e.g. Foody, 1996; Bastin, 1997; Zhang and Kirby, 1999), and the strength of class membership when classifiers based on artificial neural networks are used (e.g. Foody, 1996; Zhang and Foody, 2001; Brown *et al.* 2009). These classification methods assign, to each pixel of the image and to each class, a probability, possibility or degree of membership (depending on the considered assumptions and mathematical theories used). This may be interpreted as partial membership of the pixel to the classes (as in the case of mixed pixels or objects), a degree of similarity between what exists in the ground and the pure classes, or the uncertainty associated with the classification. In any of these cases, this additional information also reflects the classifier's difficulties in identifying which class should be assigned to the pixels, and therefore it may also be processed as uncertainty information. A number of approaches to assess the accuracy of these classifications has also been proposed (e.g. Foody, 1996; Arora and Foody, 1997; Binaghi *et al.*,1999; Woodcock and Gopal, 2000; Lewis and Brown, 2001; Oki *et al.*, 2004; Pontius and Connors, 2006). However, several of these approaches are based on generalized cross tabulation matrices from which a range of accuracy measures, with similarities to those widely used with hard classifications, may be derived (Doan and Foody, 2007).

Since, either a probability or possibility distribution is associated to each pixel, uncertainty measures may be applied to them, expressing the difficulty in assigning one class to each pixel, enabling an evaluation of the classification uncertainty and providing the user with additional valuable data to take decisions.

Uncertainty measures, among which the Shannon entropy is the most frequently used, have already been used by several authors to give an indication of the soft classification reliability (e.g. Maselli *et al.*, 1994; Zhu 1997). In this paper, it is investigated if the behaviour of probabilistic and fuzzy classifiers can be made with the uncertainty information provided by the classifiers and the uncertainty indices, instead of using confusion matrices and accuracy indices. This approach presents the advantage that no reference data is needed and, therefore, it can be used when these data are not available or are too costly to obtain. Since no reference data is necessary, this approach can be applied to all pixels inside the regions identified as representative of each class and, thus, no samples have to be chosen inside these regions, which means that the results are only influenced by the regions chosen as representative of each class. To evaluate the compatibility of the uncertainty information provided by uncertainty measures with the accuracy information, both approaches were applied to the same study area, considering two different classification methodologies. A supervised Bayesian classifier based on the maximum likelihood classifier and a supervised fuzzy classifier based on the underlying logic of Minimum-Distance-to-Means classifier were used. Several uncertainty measures were applied to estimate the classification uncertainty in the regions identified as representative of each class. Since a comparison with the results provided by the accuracy indices was necessary, these measures were only applied to the sample of pixels used to build a confusion matrix and estimate the accuracy indices, so that the results were not influenced by the representativeness of the sample used for the accuracy assessment. The results are presented and some conclusions are drawn.

## 15.2 Data and Methods

### 15.2.1 Data

The study was conducted in a rural area with a smooth topographic relief, situated in a transition zone between the centre and south of Portugal, featuring diverse landscapes representing Mediterranean environments. The area is occupied mainly by agriculture, pastures, forest and agro-forestry areas where the dominant forest species in the region are eucalyptus, coniferous and cork trees. An image obtained by the IKONOS sensor was used, with a spatial resolution of 1m in the panchromatic mode and 4m in the multi-spectral mode (XS). The product acquired was the Geo Ortho Kit and the study was performed using the four multi-spectral bands. The geometric correction of the multi-spectral image consisted of its orthorectification. The average quadratic error obtained for the geometric correction was of 1.39m, less than half the pixel size, which guarantees an accurate geo-referencing. The acquisition details of the image are presented in Table 15.1.

Table 15.1. Acquisition details of IKONOS image.

| Date | 06/04/2004 |
|---|---|
| Sun elevation (deg) | 74.8 |
| Sensor elevation (deg) | 55.5 |
| Dimension (m x m) | 11884 x 14432 |
| Bits/pixel | 11 |

### 15.2.2 Classification methods

Two pixel based classifications were applied to the IKONOS images to obtain the elementary entities, like tree crowns and parts of buildings, called Surface Elements (SE), which are the basic units of landscape used to produce a Surface Elements Map (SEM). Before the classification itself, various preliminary processing steps were carried out. First, an analysis of the image by a human interpreter was made to define the most representative classes and their surface elements. The classes used in this study are the following: Eucalyptus Trees (ET); Cork Trees (CKT), Coniferous Trees (CFT); Shadows (S); Shallow Water (SW), Deep Water (DW), Herbaceous Vegetation (HV), Sparse Herbaceous Vegetation (SHV) and Non-Vegetated Area (NVA).

Two classification methods were used: 1) a pixel-based classification of the image was performed using a supervised Bayesian classifier similar to the maximum likelihood classifier; 2) a pixel-based supervised fuzzy classifier based on the underlying logic of Minimum-Distance-to-Means classifier.

The maximum likelihood classifier is based on the estimation of the multivariate Gaussian probability density function of each class using the classes statistics (mean, variance and covariance) estimated from the sample pixels of the training set (Figure 15.1). The probability density is given by (Foody et al., 1992),

$$p(x \mid i) = \frac{1}{(2\pi)^{K/2} |V_i|^{1/2}} \exp\left(-1/2\left[\left(X - \mu_i\right)^T V_i^{-1} \left(X - \mu_i\right)\right]\right) \qquad (15.1)$$

where $p(x \mid i)$ is a probability density function of a pixel $x$ as a member of class $i$, k is the number of bands, and X is the vector denoting the spectral response of pixel $x$, $V_i$ is the variance-covariance matrix for class $i$ and $\mu_i$ is the mean vector for class $i$ over all pixels.

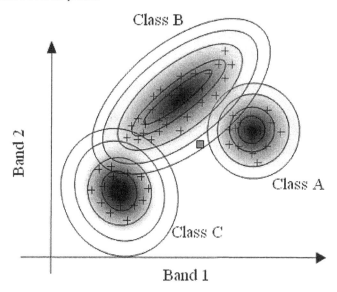

+ Values obtained for each class in the training phase

▣ Spectral response of a pixel to be classified

**Figure 15.1.** Isolines of the probability density functions associated to classes A, B and C.

The traditional use of this classification method assigns each pixel to the class corresponding to the highest probability density function value. However, the posterior probabilities can be computed with the probability density functions. The posterior probabilities of a pixel $x$ belonging to class i, denoted by $p_i(x)$, is given by Equation 15.2,

$$p_i(x) = \frac{P(i)p(x \mid i)}{\sum_{i=1}^{n} P(i)p(x \mid i)} \qquad (15.2)$$

where $P(i)$ is prior probability of class $i$ (the probability of the hypothesis being true regardless of the evidence) and $n$ is the number of classes. These posterior probabilities sum up to one for each pixel (Foody et al., 1992), and may be interpreted as representing the proportional cover of the classes in each pixel or as indicators of the uncertainty associated with the pixel allocation to the classes (e.g. Shi et al., 1999; Ibrahim et al., 2005). The second interpretation is used in this paper, where the posterior probabilities are used to compute uncertainty measures.

The second classification method used is a pixel-based supervised fuzzy classifier based on the underlying logic of Minimum-Distance-to-Means Classifier (FMDMC), available in the commercial software IDRISI. With this method, an image is classified based on the information contained in a series of signature files and a standard deviation unit (Z-score distance) introduced by the user. The fuzzy set membership is calculated based on a standardized Euclidean distance from each pixel reflectance, on each band, to the mean reflectance for each class signature (see Figure 15.2), using a Sigmoidal membership function (see Figure 15.3). The underlying logic is that the mean of a signature represents the ideal point for the class, where fuzzy set membership is one. When distance increases, fuzzy set membership decreases, until it reaches the user-defined Z-score distance where fuzzy set membership decreases to zero. To determine the value to use for the standard deviation unit, the information of the training data set was used to study the spectral separability of the classes and determine the average separability measure of the classes.

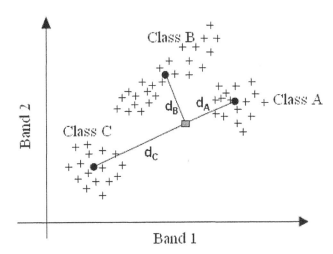

+ Values obtained for each class in the training phase
● Mean of the values obtained in the training phase
▣ Spectral response of a pixel to be classified

**Figure 15.2.** Minimum-Distance-to-Means Classifier.

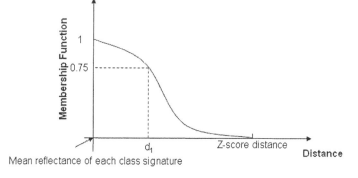

**Figure 15.3.** Membership function used in the fuzzy classifier based on the Minimum-Distance-to-Means Classifier.

With this classification methodology the sum of the degrees of membership of each pixel to each class may sum up to any value between zero and the number of classes. Since each fuzzy set induces a possibility distribution, a possibility distribution associated to each pixel is obtained.

Unlike traditional hard classifiers, the output obtained with these classifiers is not a single classified map, but rather a set of images (one per class) that express, for the first classifier, the probability, and for the second one, the possibility that each pixel belongs to the class in question.

Both classification methods require a training stage, where a training set is chosen to obtain class descriptors. The training dataset consisted of a semi-random selection of sites. A human interpreter delimited twenty-five polygons for each class and a stratified random selection of 300 samples per class was performed. One half of these points was used to train the classifier, and the other half presented a spectral variability in class response very similar to the training set and was used to evaluate the classifier's performance. The sample unit was the pixel and only pure pixels were included in both samples (Plantier and Caetano, 2007). The total sample size used for training included 1350 pixels. An equal number of pixels was used to test the classifier accuracy. The total number of pixels used corresponds to 5% of the pixels inside the chosen polygons.

### 15.2.3 Classifiers evaluation with the accuracy assessment

The accuracy assessment was made with an error matrix. The reference data were obtained from aerial images with larger resolution. The Global Accuracy was computed, as well as the Khat coefficient. The User Accuracy (UA) and the Producer Accuracy (PA) indices were also computed for all classes.

### 15.2.4 Classifiers evaluation with uncertainty measures

Uncertainty measures were used to quantify the uncertainty of both classifications. An overview of the uncertainty measures used follows.

### 15.2.4.1 IDRISI uncertainty measure

The uncertainty measure provided in IDRISI is a ratio computed using Equation 15.3. This uncertainty measure can be applied to probability and possibility distributions, and therefore $p_i$ (i=1,...,n) represents the probabilities or possibilities associated with the several classes and n is the number of classes (Eastman, 2006),

$$RI = 1 - \frac{\max_{i=1,..n}(p_i) - \frac{\sum_{i=1}^{n} p_i}{n}}{1 - \frac{1}{n}} \qquad (15.3)$$

This uncertainty measure assumes values in the interval [0,1]. When applied to probability distributions the sum of the probabilities have to add up to one and the measure only depends on the maximum probability and the total number of classes. When applied to possibility distributions, the sum of the possibilities can vary between 0 and $n$. The numerator of the second term expresses the difference between the maximum probability or possibility assigned to a class and the probability or possibility that would be associated with the classes if a total dispersion for all classes occurred. The denominator expresses the extreme case of the numerator, where the maximum probability or possibility is one (and thus a total commitment to a single class occurs). The ratio of these two quantities expresses the degree of commitment to a specific class relative to the largest possible commitment (Eastman, 2006). The classification uncertainty is thus the complement of this ratio and evaluates the degree of compatibility with the most probable or possible class and until at which point the classification is dispersed over more than one class, providing information regarding the classifier's difficulty in assigning only one class to each pixel.

### 15.2.4.2 Entropy

The Shannon entropy measure ($E$), derived from Shannon's information theory, is computed using Equation 15.4 (Ricotta, 2005).

$$E = -\sum_{i=1}^{n} p_i \log_2 p_i \qquad (15.4)$$

This measure $E$ assumes values in the interval [0, $\log_2 n$]. Since a comparison of the results given by several measures will be done in this study, it is necessary to be able to do a direct comparison between the results given by them, which is not possible if the measures used have different ranges. Therefore, to overcome this problem, a measure of the relative probability entropy (RPH) was used, similar to the one proposed by Maselli *et al.* (1994), and given by Equation 15.5,

$$RPH = \frac{-\sum_{i=1}^{n} p_i \log_2 p_i}{\log_2 n} \qquad (15.5)$$

RI and RPH both vary between zero and one, meaning, for the value of zero there is no ambiguity in the assignment of the pixel to a class, and therefore no uncertainty, and for the value of one, the ambiguity is maximum.

### 15.2.4.3 Non-specificity measures

Non-specificity measures have been developed within the context of possibility theory and fuzzy sets theory, namely, the *NSp* non-specificity measure, proposed by Yager (1982); and the U-uncertainty measure, proposed by Higashi and Klir (1982). Both measures can be applied to fuzzy sets and to ordered possibility distributions. The non-specificity measures are appropriate to evaluate the uncertainty resulting from possibilistic classifications, since they quantify the ambiguity in specifying an exact solution (e.g. Pal and Bezdek, 2000).

For an ordered possibility distribution $\Pi$ defined over a universal set X, that is, a possibility distribution such that $\Pi(x_1) \geq \Pi(x_2) \geq ... \geq \Pi(x_n)$, the *NSp* non-specificity measure is given by Equation 15.6,

$$NSp(\Pi) = 1 - \sum_{i=1}^{n}\left[\Pi(x_i) - \Pi(x_{i+1})\right]\frac{1}{i} \qquad (15.6)$$

where n is the number of elements in the universal set and $\Pi(x_{n+1})$ is assumed to be zero. The U-uncertainty measure, proposed by Higashi and Klir (1982), applied to an ordered possibility distribution $\Pi$, is given by Equation 15.7,

$$U(\Pi) = \int_{0}^{\alpha_{max}} \log_2 |\Pi_\alpha| d\alpha + (1 - \alpha_{max})\log_2 n \qquad (15.7)$$

where $\Pi_\alpha$ is the $\alpha$-possibility level of $\Pi$, which corresponds to the subset of elements having possibility at least $\alpha$, that, in mathematical terms, can be expressed as $\Pi_\alpha = \{x | \Pi(x) \geq \alpha\}$; where $|\Pi_\alpha|$ is the cardinality of $\Pi_\alpha$, that is, the number of elements of $\Pi_\alpha$; and $\alpha_{max}$ is the largest value of $\Pi(x)$ for every element of the universal set X, which can be expressed as $\alpha_{max} = \max_x \Pi(x)$, $\forall x \in X$.

An approximate solution of Equation 15.7 is (Mackay *et al.*, 2003)

$$U(\Pi) = \left[1 - \Pi(x_1)\right]\log_2 n + \sum_{i=2}^{n}\left[\Pi(x_i) - \Pi(x_{i+1})\right]\log_2 i \qquad (15.8)$$

where $\Pi(x_{n+1})$ is assumed to be zero.

Properties of both measures can be found in Pal and Bezdek (2000) or Klir (2000). From these properties we stress the following:

- For a possibility distribution $\Pi$, $0 \leq NSp(\Pi) \leq 1$, where the minimum is obtained for $\Pi(x_1, x_2, ..., x_n) = (1, 0, ..., 0)$ and the maximum for $\Pi(x_1, x_2, ..., x_n) = (0, 0, ..., 0)$;
- For a universal set X and a possibility distribution $\Pi$, $0 \leq U(\Pi) \leq \log_2 n$, where the minimum is obtained for $\Pi(x_1, x_2, ..., x_n) = (1, 0, ..., 0)$ and the maximum value for $\Pi(x_1, x_2, ..., x_n) = (\alpha, \alpha, ..., \alpha)$, $\forall \alpha \in [0,1]$.

Since both measures have different ranges, and a comparison of the results given by both will be made in this paper, a normalized version proposed by Gonçalves *et al.* (in press) of the U-uncertainty measure is used, varying within the interval [0,1] and given by Equation 15.9.

$$U_n(\Pi) = \frac{[1 - \Pi(x_1)]\log_2 n + \sum_{i=2}^{n}[\Pi(x_i) - \Pi(x_{i+1})]\log_2 i}{\log_2 n} \qquad (15.9)$$

To evaluate the uncertainty of the probabilistic classifier, the uncertainty measure *RI* (available in the software IDRISI) and the *RPH* were used. To evaluate the uncertainty of the possibilistic classifier, the two non-specificity measures *Un* and *NSp* were used, as well as the uncertainty measure RI.

Even though uncertainty measures can be computed to all pixels of an image, since the objective was to compare the results given by the uncertainty measures with the ones given by the error matrix for points in the sample used to test the classifier, only these pixels were used in this analysis.

## 15.3 Results and Discussion

### 15.3.1 Classifiers evaluation with accuracy indices

Table 15.2 shows the results obtained for the Global Accuracy and the Khat Coefficient obtained for both classifications, and Tables 15.3 and 15.4 presents the confusion matrixes built.

**Table 15.2.** Accuracy indices for the fuzzy and Bayes classifiers

|  | Global Accuracy (%) | KHAT Coefficient (%) |
|---|---|---|
| **Bayes** | 96.68 | 96.25 |
| **Fuzzy** | 84.54 | 82.55 |

**Table 15.3.** Confusion matrices for the Fuzzy classifier.

|  | DW | SW | NVA | ET | S | HV | CKT | CFT | SHV | UA |
|---|---|---|---|---|---|---|---|---|---|---|
| **DW** | 153 |  |  |  | 2 |  |  |  |  | 98.7 |
| **SW** |  | 45 | 4 |  |  |  |  |  |  | 91.8 |
| **NVA** | 16 | 28 | 150 | 5 | 4 | 15 | 7 | 15 | 22 | 57.3 |
| **ET** |  |  |  | 118 |  |  | 2 | 19 |  | 84.9 |
| **S** |  |  |  |  | 131 |  |  |  |  | 100.0 |
| **HV** |  |  |  |  |  | 173 |  | 1 |  | 99.4 |
| **CKT** |  |  |  | 2 |  |  | 145 |  | 20 | 86.8 |
| **CFT** |  |  |  | 14 |  |  |  | 130 |  | 90.3 |
| **SHV** |  |  |  |  |  |  | 18 |  | 103 | 85.1 |
| **PA** | 90.5 | 61.6 | 97.4 | 84.9 | 95.6 | 92.0 | 84.3 | 78.8 | 71.0 | **84.54%** |

**Table 15.4.** Confusion matrices for the Bayes classifier.

|  | DW | SW | NVA | ET | S | HV | CKT | CFT | SHV | UA |
|---|---|---|---|---|---|---|---|---|---|---|
| **DW** | 166 |  | 1 |  |  |  |  |  |  | 99.4 |
| **SW** | 1 | 74 |  |  |  |  |  |  |  | 98.7 |
| **NVA** | 1 |  | 177 |  | 1 |  |  |  |  | 98.9 |
| **ET** |  |  |  | 130 |  |  |  | 3 |  | 97.7 |
| **S** | 1 |  | 1 |  | 138 |  |  |  |  | 98.6 |
| **HV** |  |  |  |  |  | 201 |  |  |  | 100.0 |
| **CKT** |  |  | 1 | 1 |  |  | 160 |  | 14 | 90.9 |
| **CFT** |  |  |  | 8 |  |  |  | 162 |  | 95.3 |
| **SHV** |  |  |  |  |  |  | 12 |  | 131 | 91.6 |
| **PA** | 98.2 | 100.0 | 98.3 | 93.5 | 99.3 | 100.0 | 93.0 | 98.2 | 90.3 | **96.75%** |

According to both accuracy indices, the two classifiers present good results, but both indices present better results for the Bayes classifier. Figures 15.4 and 15.5 show the results obtained for the UA and PA indices for both classifiers, ordered with increasing values for the fuzzy classifier.

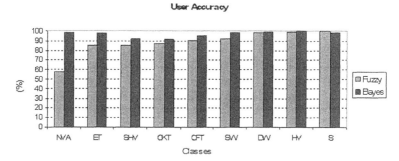

**Figure 15.4.** User Accuracy computed for all classes for the Fuzzy and Bayes classifiers.

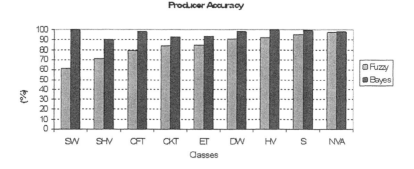

**Figure 15.5.** Producer Accuracy computed for all classes for the Fuzzy and Bayes classifiers.

These Figures show that, for the Bayes classifier, the UA and the PA have values larger than 90% for all classes. The classes with lower accuracy are CKT and SHV. Since the accuracy indices presented very high values for all classes it may be concluded that the classifier has a good performance in their identification.

The results obtained for the fuzzy classifier are worse. Only the classes HV, S and DW have UA and PA larger than 90%. Moreover, all classes except the S class present lower values of UA and PA than the Bayes classifier. The class with the smaller value of PA is SW (62%), which means it is the class with more omission errors. The pixels that are erroneously not included in this class are included in the NVA class, which is the class with smaller UA (57%) and therefore the class with more commission error, receiving pixels from all classes. This shows that this classifier presents great difficulty in classifying the NVA and SW classes.

### 15.3.2 Classifiers evaluation with uncertainty indices

Figure 15.6 shows the mean uncertainty per class obtained with the *RI* and *En* uncertainty indices for the probabilistic classification, ordered with increasing values of *RI*. Figure 15.7 shows the uncertainty values grouped into five levels, corresponding to 0% and to quartiles.

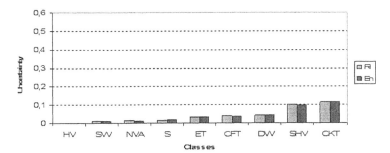

**Figure 15.6.** Mean uncertainty per class obtained for the test sample with the uncertainty measures *En* and *RI* considering the Bayes classifier.

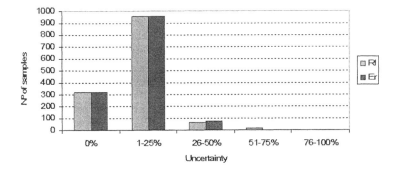

**Figure 15.7.** Distribution of the values obtained for the *RI* and *En* uncertainty measures for the Bayes classifier considering five levels of uncertainty.

Figures 15.6 and 15.7 show that the results obtained with both measures are almost identical. The values of the mean uncertainty per class are very low, always less than 0.12. Figure 15.7 shows that the majority of samples have uncertainty values in the interval 1-25%. Figure 15.6 shows that the classes showing higher levels of uncertainty are CKT and SHV. The information provided by the uncertainty measures is therefore in accordance with the one provided by the accuracy indices. Figure 15.8 shows the mean uncertainty per class obtained with the *NSp*, *Un* and *RI* uncertainty measures applied to the fuzzy classification, ordered with increasing values of *NSp*. Figure 15.9 shows the uncertainty values grouped into five levels, corresponding to 0% and to quartiles.

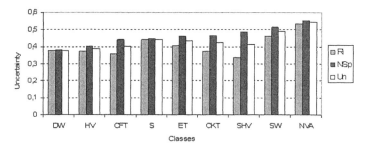

**Figure 15.8.** Mean uncertainty per class obtained for the test sample with the uncertainty measures *RI*, *NSp* and *Un* considering the fuzzy classifier.

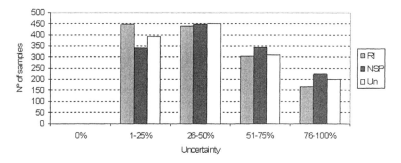

**Figure 15.9.** Distribution of the values obtained for the *RI*, *NSp* and *Un* uncertainty measures for the fuzzy classifier considering five levels of uncertainty.

The values of the mean uncertainty obtained with the uncertainty measures *RI*, *Nsp* and *Un* for the fuzzy classification, shown in Figure 15.9, vary between 0.3 and 0.55. The values obtained with the several uncertainty measures are in general very close, although the *NSp* values are in general slightly higher than the ones obtained with the other measures. Small discrepancies can be observed between the uncertainty values obtained with the three uncertainty measures, which are explained by the fact that they evaluate different characteristics of the possibility distributions (for more information see Gonçalves *et al.*, 2008 and Gonçalves *et al.*, in press). However, all measures present higher uncertainty levels for the NVA and SW classes. Since with this classifier the uncertainty values are relatively high, an additional analysis of the possibility distributions associated with the pixels of the test sample was made, to understand why the uncertainty values were so large. This analysis showed that, for all classes, there were a large percentage of pixels with relatively low values for the largest degree of possibility. For example, 25% of the sample pixels assigned to the DW class had the largest degree of possibility smaller than 0.5, which explains the low values of certainty. Moreover, no pixel of the testing sample had the largest possibility equal to 1 (maximum possibility).

For all classes, except NVA, the second degree of possibility was assigned to the NVA class with degrees of possibility frequently very close to the maximum. This means that there is great confusion between NVA and the other classes, which is one of the main problems of the application of this classifier to this case study. This conclusion was also obtained from the accuracy indices, since the NVA class presented a very large difference between the UA (57.3%) and the PA (97.4%). This means that this class presents a large number of commission errors and, therefore, samples that were assigned to the NVA class should have been assigned to other classes as can be seen in the error matrix (see Figure 15.3), such as SW and SHV, explaining why they have larger omission errors.

## 15.4 Conclusions

The results obtained with both classifiers show that the information about the uncertainty of the classification of regions considered as representative of the several classes can be used, together with the degrees of probability or possibility assigned to each pixel, to detect the main problems of the classifiers. For both classifiers used, the same conclusions were reached about the performance of the classifiers with the accuracy indices and the uncertainty information. Since it is very simple to apply the uncertainty measures to the probability or possibility distributions, with this approach it is much faster and cheaper to evaluate the classifiers' performance than with the accuracy indices. Also, unlike the accuracy results, they are not influenced by the subjectivity of the reference data. Besides that, they can be applied to all pixels of the regions chosen as representative of the several classes, since no reference information is necessary. For all these reasons it is a promising approach, which provides valuable information to the user, and may reduce time and costs, therefore deserving further attention.

## References

Arora, M.K. and G.M. Foody. (1997), "Log-linear modelling for the evaluation of the variables affecting the accuracy of probabilistic, fuzzy and neural network classifications". *International Journal of Remote Sensing*, Vol. 18 (4): 785-798.

Bastin, L. (1997), "Comparison of fuzzy c-means classification, linear mixture modeling and MLC probabilities as tools for unmixing coarse pixels". *International Journal of Remote Sensing*, Vol. 18 (17): 3629-3648.

Binaghi, E., P.A. Brivio, P. Ghezzi and A. Rampini. (1999), "A fuzzy set-based accuracy assessment of soft classification". *Pattern Recognition Letters*, Vol. 20 (9): 935-948.

Brown, K.M., G.M. Foody and P.M. Atkinson. (2009), "Estimating per-pixel thematic uncertainty in remote sensing classifications". *International Journal of Remote Sensing*, Vol. 30 (1): 209-229.

Doan, H.T.X. and G.M. Foody. (2007), "Increasing soft classification accuracy through the use of an ensemble of classifiers". *International Journal of Remote Sensing*, Vol. 28 (20): 4609-4623.

Eastman, J.R. (2006), *IDRISI Andes Guide to GIS and Image Processing*. Clark Labs., Clark University, 327p.

Foody, G.M., N.A. Campbell, N.M. Trodd and T.F. Wood. (1992), "Derivation and applications of probabilistic measures of class membership from maximum likelihood classification". *Photogrammetric Engineering and Remote Sensing*, Vol. 58 (9): 1335-1341.

Foody, G.M. (1996), "Approaches for the production and evaluation of fuzzy land cover classifications from remotely-sensed data". *International Journal of Remote Sensing*, Vol. 17 (7): 1317-1340.

Gonçalves, L.M.G., C.C. Fonte, E.N.B.S. Júlio and M. Caetano. (2008), "A method to incorporate uncertainty in the classification of remote sensing images". *In:* J. Zhang and M.F. Goodchild (eds.) *Proceedings of 8th International symposium on spatial accuracy assessment in natural resources and environmental sciences (Accuracy 2008)*, Shanghai, China, pp. 179-185.

Gonçalves, L.M.G., C.C. Fonte, E.N.B.S. Júlio and M. Caetano. (in press), "Evaluation of soft possibilistic classifications with non-specificity uncertainty measures". *International Journal of Remote Sensing*.

Gopal, S. and C.E. Woodcock. (1994), "Theory and methods for accuracy assessment of thematic maps using fuzzy sets". *Photogrammetric Engineering and Remote Sensing*, Vol. 60 (2): 181-188.

Higashi, M. and Klir, G., 1982, Measures of uncertainty and information based on possibility distributions. *International Journal of General Systems*, Vol. 9 (1): 43-58.

Ibrahim, M.A., M.K. Arora and S.K. Ghosh. (2005), "Estimating and accommodating uncertainty through the soft classification of remote sensing data". *International Journal of Remote Sensing*, Vol. 26 (14): 2995-3007.

Klir, G. (2000), "Measures of uncertainty and information". *In:* D. Dubois and H. Prade (eds.) *Fundamentals of Fuzzy Sets*, (The Handbook of Fuzzy Sets Series, Vol. 7), Dordrecht, Kluwer Academic Publishing, pp. 439-457.

Lewis, H.G. and M. Brown. (2001), "A generalized confusion matrix for assessing area estimates from remotely sensed data". *International Journal of Remote Sensing*, Vol. 22 (16): 3223-3235.

Mackay, D., S. Samanta, D. Ahl, B. Ewers, S. Gower and S. Burrows. (2003), "Automated Parameterization of Land Surface Process Models Using Fuzzy Logic". *Transactions in GIS*, Vol. 7 (1): 139-153.

Maselli, F., C. Conese and L. Petkov. (1994), "Use of probability entropy for the estimation and graphical representation of the accuracy of maximum likelihood classifications". *ISPRS Journal of Photogrammetry and Remote Sensing*, 49 (2): 13-20.

Oki, K., T.M. Uenishi, K. Omasa and M. Tamura. (2004), "Accuracy of land cover area estimated from coarse spatial resolution images using an unmixing method". *International Journal of Remote Sensing*, Vol. 25 (9): 1673-1683.

Pal, N. and J. Bezdek, (2000), "Quantifying different facets of fuzzy uncertainty". *In:* D. Dubois, and H. Prade (eds.) *Fundamentals of Fuzzy Sets*, (The Handbook of Fuzzy Sets Series, Vol. 7), Dordrecht, Kluwer Academic Publishing, pp. 459-480.

Plantier, T. and M. Caetano. (2007), "Mapas do Coberto Florestal: Abordagem Combinada Pixel/Objecto". *In:* J. Casaca and J. Matos (eds.) *V Conferência Nacional de Cartografia e Geodesia*, Lisboa, Lidel, pp. 157-166.

Pontius, R.G. and J. Connors. (2006), "Expanding the conceptual, mathematical and practical methods for map comparison". *In:* M. Caetano and M. Painho (eds.) *Proceedings of 8th International symposium on spatial accuracy assessment in natural resources and environmental sciences (Accuracy 2006)*, Lisbon, Portugal, pp. 64-79.

Ricotta, C. (2005), "On possible measures for evaluating the degree of uncertainty of fuzzy thematic maps". *International Journal of Remote Sensing*, Vol. 26 (24): 5573-5583.

Shi, W.Z., M. Ehlers and K. Tempfli. (1999), "Analytical modelling of positional and thematic uncertainties in the integration of remote sensing and geographical information systems". *Transactions in GIS*, Vol. 3 (2): 119-136.

Stehman, S.V. and R.L. Czaplewski. (2003), "Introduction to special issue on map accuracy". *Environmental and Ecological Statistics*, Vol. 10 (3): 301-308.

Yager, R. (1982), "Measuring tranquility and anxiety in decision making: an application of fuzzy sets". *International Journal of General Systems*, Vol. 8 (3): 139-146.

Woodcock, C.E. and S. Gopal. (2000), "Fuzzy set theory and thematic maps: accuracy assessment and area estimation". *International Journal of Geographical Information Science*, Vol. 14 (2): 153-172.

Zhang, J. and G.M. Foody. (2001), "Fully-fuzzy supervised classification of sub-urban land cover from remotely sensed imagery: statistical and artificial neural network approaches". *International Journal of Remote Sensing*, Vol. 22 (4): 615-628.

Zhang, J. and R.P. Kirb. (1999), "Alternative Criteria for Defining Fuzzy Boundaries Based on Fuzzy Classification of Aerial Photographs and Satellite Images". *Photogrammetric Engineering and Remote Sensing*, Vol. 65 (12): 1379-1387.

Zhu, A. (1997), "Measuring uncertainty in class assignment for natural resource maps under fuzzy logic". *Photogrammetric Engineering and Remote Sensing*, Vol. 63 (10): 1195-1202.

# 16

## Short Notes

## Analyzing the Effect of the Modifiable Areal Unit Problem in Remote Sensing

*Nicholas A.S. Hamm, Alfred Stein & Valentyn Tolpekin*

The modifiable areal unit problem (MAUP) has long been recognized as a confounding factor in GIS analysis. However, it has received relatively little attention in remote sensing. The MAUP has two components: scale and zonation. The scale component recognizes that examining different scales of aggregation may show different patterns in the data. In remote sensing, different scales might correspond to different pixels sizes (e.g., 1m, 3m, 5m, 15m, 30m). The zonation component recognizes that boundaries can be located in different places at any given scale. This is most obvious in vector GIS, where multiple polygons of the same area, but with differing shapes, can be formed. In remote sensing, the zonation component is apparent when we consider the multiple possible orientations of the image grid with the scene. Different zonation structures at the same scale may reveal different patterns in the data. The scale and zonation components of the MAUP are both important in remote sensing; however, the zonation component has received little attention.

In this research, we began by examining the effect of the MAUP on the classification of a simple, simulated, 2-class image (lake and land). This was examined at multiple pixel resolutions from a 1m base pixel size up to 10m aggregated pixel. The scale component was examined first. Large variability in the size of the classified lake was observed ($294m^2$ to $512m^2$), but this had no relationship with the pixel size. For each level of aggregation, multiple different zonation structures were realized by shifting the image grid relative to the scene. This also showed wide variation in the estimate of the area of the lake. For example, for a 3m aggregated pixel, the estimated area of the lake ranged from 423 to $468m^2$. This analysis of the impact of zonation on image classification is interesting and important because it shows that, all other things being equal, the arbitrary location of the image relative to the scene can have profound effects on inferences drawn from the image.

This research is currently being extended in two ways. First, we are extending it to analyze a more complex 4-class, multispectral simulated image. Extending the research with a simulated image is valuable, because it allows us to control the pixel size, zonation and class separability. The second extension focuses on the classification of a 'real' image for a simple landscape. This will allow us to examine the extent to which zonation effects are present in thematic maps classified from real remotely sensed imagery. The impacts of the point spread function and of different re-sampling techniques are also important areas of research.

# Analyzing the Effect of Different Aggregation Approaches on Remotely Sensed Data

*Rahul Raj, Nicholas A.S. Hamm & Yogesh Kant*

Spatial aggregation is widely used in land-use and land-cover mapping and monitoring, and ecological resource management, which are all carried out at local, national, regional and global scales. Categorical aggregation (spatial aggregation of classified images) and numerical aggregation (spatial aggregation of continuous images) have different impacts on the inferences drawn from aggregated images. This research examined the effect of both aggregation approaches, since understanding their effects can help to select the appropriate method for a particular study. Under categorical aggregation, the effect of the majority rule-based (MRB), random rule-based (RRB) and point-centred distance-weighted moving window (PDW) methods were analyzed (Gardner *et al.*, 2008). Under numerical aggregation, the effect of adopting the mean and of performing central pixel resampling was examined. Most of the previous studies on the effect of aggregation were carried out using land-cover classes. The present study evaluated the effect of aggregation using land-use/cover classes in order to expand the generality of the results. A LISS-III image (23.5m) was aggregated to 70.5m, 117.5m and 164.5m before and after classifying it.

The change in class proportions, aggregation index (AI), and square pixel index (SqP) were used to assess the effect of aggregation approaches with respect to the reference fine-resolution image. AI gives a measure of how clumped different classes are (He *et al.*, 2000), and SqP is a measure of shape complexity (Frohn, 1997). The class area changes with aggregation level. Hence, a calibration-based model was applied to correct the area of classes with respect to the reference area. The effect of the MRB and mean approaches on the calibration-based model was also addressed. The local variance of each class was also computed to assess the effect of numerical aggregation approaches. The RRB approach is based on random selection of the class from an input grid, so it produces different realizations of aggregated images for a given fine-resolution input. The PDW approach is based on selection of different parameters values, so changing these parameters leads to changes in the output image. The variability in RRB and PDW were reported by assessing their effect on class proportions, AI and SqP value.

Results indicated that the RRB, PDW and CPR approaches preserved the class proportion with decreasing spatial resolutions. MRB increased the proportion of the dominant class and decreased all other class proportions. On the contrary, the mean approach increased the proportion of classes that were not dominant in the landscape. MRB increased the AI value of the dominant class and decreased the AI value of other classes. RRB, PDW and CPR decreased the AI value of all classes, but the rate of decrease was less for the mean aggregation approach. The RRB, PDW and CPR approaches led to low distortion in shape complexity, as measured by SqP, of all classes relative to the MRB and mean approaches. Hence, they better preserved the complexity of spatial information. Since RRB, PDW and CPR maintained the class proportions the derivation of a calibration based model was not necessary. Where the calibration model was applied, MRB gave more accurate results than the mean approach. The graph of local variance showed that mean approach decreased the local variance of each class, whereas the CPR approach led to a strong increase in local variance. For the different realizations, RRB responded in similar manner. For the PDW method most parameters should be fixed according to the level of aggregation. The exception is the weight parameter, $w$. The local variance can be minimized by setting $w=2$.

# References

Frohn, R.C. (1997), *Remote Sensing for Landscape Ecology*, Lewis Publishers, Boca Raton, 99p.

Gardner, R.H., T.R. Lookingbill, P.A. Townsend and J. Ferrari. (2008), "A new approach for rescaling land cover data". *Landscape Ecology*, Vol. 23: 513-526.

He, H.S., B.E. DeZonia and D.J. Mladenoff. (2000), "An aggregation index (AI) to quantify spatial patterns of landscapes". *Landscape Ecology*, Vol. 15: 591-601.

# Use of an Airborne Imaging Spectrometer as a Transfer Standard for Atmospheric Correction of SPOT-HRG Data

*Debajyoti Bhowmick, Nicholas A.S. Hamm & Edward J. Milton*

Validated and quality assured Earth observation (EO) data products are required for environmental monitoring and for use in the various climatological and ecosystem models that are essential for the advance of Earth system science. Furthermore, remotely sensed data are often used in combination with in-situ measurements for long term monitoring and modelling. Validation of remotely sensed data products requires establishing relationships between field measurements and imagery through data aggregation and up-scaling. This up-scaling is non-trivial when the very different scales are considered. For example, a field measurement may have a spatial support of a few centimetres, whereas the pixel-size in a coarse resolution image may be several kilometres. In order to address this issue, this study uses an airborne image as a "transfer-standard". The application under consideration is the atmospheric correction of optical remotely sensed data and used the Specim AISA-Eagle (1m pixels) imaging spectrometer as a transfer standard towards atmospheric correction of SPOT-HRG (10m pixels) data. The objective was to provide a correction that is traceable to a validated data product with known uncertainty. All data were obtained from the UK Network for Calibration and Validation of Earth Observation (NCAVEO) field campaign run in June 2006 (Milton and NCAVEO Partnership, 2006). This field campaign obtained a unique and wide-ranging dataset which included field data, airborne imagery and satellite imagery. The data used in this project include field measurements of spectral reflectance, airborne hyperspectral imagery (Specim AISI-Eagle, CASI (compact airborne imaging spectrometer) 2, CASI 3) and SPOT-HRG, and were all obtained within 24 hours. Furthermore, the Eagle and CASI 2 instruments were mounted next to each other on the same aircraft. This allowed assessment of the Eagle against the CASI 2 instrument. This is of immediate practical benefit since the Eagle is relatively new and untested, whereas the CASI is known to be stable and accurately calibrated.

Airborne imaging spectrometers are able to identify and quantify atmospheric absorption features and scattering curves, allowing correction of spectra to values of reflectance. Clark *et al.* (1995) proposed a hybrid approach called RTGC (Radiative Transfer Ground Calibration), which used a radiative transfer (RT) model and spectra gathered from ground calibration targets. The RT model corrects the image based on spectral absorption features and the spectra from the ground targets are used to remove the residual errors commonly associated with RT models. This approach was applied to the Eagle data, using atmospheric data and field spectra obtained during the NCAVEO campaign, to obtain a fine-resolution hyperspectral reflectance map. The reflectance map (the transfer standard) was then validated using two different approaches, (1) direct comparison with separate field-measured ground targets, and (2) an indirect method that compared the reflectance map to an independent map produced from the CASI sensor. The rationale for this second validation was that the CASI sensor is known to be stable and accurately calibrated. Because a reflectance map is a standardized product, if comparable results are derived from two standardized products it points towards a calibrated product and can be used with confidence. A similar approach is adopted for MODIS and MERIS products. The CASI and Eagle reflectance maps were shown to agree to within 8%. Radiometric linearity between digital number (DN) and radiance was observed for the AISA-Eagle, and the reflectance map generated was in close match with surface reflectance (within 6%), establishing the use of Eagle as a potential transfer function. The generated reflectance map with known uncertainty was then used to atmospherically correct SPOT-HRG in order to derive a validated and traceable SPOT reflectance map.

Atmospheric correction of the SPOT data was performed using the empirical line method (ELM). This is based on a linear relationship between radiance and reflectance and requires estimation of the slope and intercept coefficients of a regression line. Data for the regression model were obtained using two approaches. First, a small number of targets (3-5) were obtained in both the Eagle (reflectance) and SPOT (radiance) images. The values within those targets were averaged and then used in the regression model. This led to an uncertain regression model, as illustrated by the confidence intervals on the regression line (up to 5% reflectance). The second approach was based on a pixel-matching between the two images, yielding several hundred points for use in the regression model. This required minimal additional effort and led to substantial increase in the confidence intervals on the regression line (up to 2% reflectance). It also opens the possibility for more advanced regression modelling which would take into account both spatial and spectral autocorrelation.

## References

Milton, E.J. and NCAVEO Partnership. (2008), "The Network for Calibration and Validation in Earth Observation (NCAVEO) 2006 field campaign". *In:* K. Anderson (ed.) *Proceedings of the Remote Sensing and Photogrammetry Society Conference: "Measuring Change in the Earth System".* University of Exeter, 15-17 September 2008. Nottingham, UK, Remote Sensing & Photogrammetry Society, pp. 1-4.

Clark, R.N., G.A. Swayze, K.B. Heidebrecht, R.O. Green and A.F.H. Goetz. (1995), "Calibration to surface reflectance of terrestrial imaging spectrometry data: Comparison of methods". *In:* R.O. Green (Ed.), *Summaries of the Fifth Annual JPL Airborne Earth Science Workshop,* JPL Publ. 95-1, Jet Propulsion Laboratory, Pasadena, CA, pp. 41-42.

# SECTION IV – LEGAL ASPECTS

# 17

## Liability for Spatial Data Quality

*Harlan J. Onsrud*

### Abstract

*Liability in data, products, and services related to geographic information systems, spatial data infrastructure, location based services and web mapping services, is complicated by the complexities and uncertainties in liability for information system products and services generally, as well as by legal theory uncertainties surrounding liability for maps. Each application of geospatial technologies to a specific use may require integration of different types of data from multiple sources, assessment of attributes, adherence to accuracy and fitness-for-use requirements, and selection from among different analytical processing methods. All of these actions may be fraught with possible misjudgments and errors. A variety of software programs may be run against a single geographic database, while a wide range of users may have very different use objectives. The complexity of the legal questions surrounding liability for geospatial data, combined with the diversity of problems to which geospatial data and technologies may be applied and the continually changing technological environment, have created unsettling and often unclear concerns over liability for geospatial technology development and use. This article selects a single data quality issue to illustrate that liability exposure. In regard to that issue, it may have a substantial stifling effect on the widespread use of web-based geospatial technologies for such purposes as geographic data mining and interoperable web mapping services. The article concludes with a recommendation for a potential web-wide community solution for substantially reducing the liability exposure of geospatial technology and geographic data producers and users.*

### 17.1 Introduction

Use of digital spatial data, unlike use of a music or journal article file, may often result in some action or decision. If errors or other shortcomings such as lack of data accuracy or completeness result in an inappropriate action or wrong decision, the possibility of liability arises for data suppliers as well as for all other parties in the chain of handling and processing the spatial data.

"As a general proposition, legal liability for damages is a harm-based concept. For instance, those who have been specifically hired to provide data for a database, or those who are offering data for sale to others, are responsible for some level of competence in the performance of the service or for some level of fitness in the product offered. If others are damaged by mistakes that a producer should not have made, or by inadequacies that should not have been allowed, the courts have reasoned that producers should bear some responsibility for the damages. But for their mistake or defective product, the damages would not have occurred. In commercial settings, liability exposure often may be reduced through appropriate communications, contracts, and business practices. However, liability exposure may never be eliminated completely. Nor as a society would we want it to be. Modern societies generally support the proposition that individuals and businesses should take responsibility for their actions if those actions have unjustifiably caused harm to others.

However, the law does not require perfection. The law exists and responds to a realistic world. No general purpose dataset will ever be complete for all potential purposes that users might desire. Nor will the accuracy of data ever meet the needs of all conceivable uses. It is also inevitable that errors and blunders will be contained in any practical database. Thus, the law holds that those in the information chain should be liable only for those damages they had a duty to prevent. Establishing the nature and extent of rightful duties has traditionally been accomplished under theories of tort or contract law. Legislation may impose additional or alternative liability burdens. Legislation affecting liability for spatial datasets and software typically might be found in statutes addressing such issues as intellectual property rights, privacy rights, anti-trust issues, and access rights" (Onsrud, 1999).

Although the facts or legal issues may be complex in a specific dispute, the core legal concepts in imposing liability on map- or mapping system creators for inaccuracies or blunders that should not have occurred and that have caused harm have not changed substantially as societal uses have moved from paper to digital formats. For an overview discussion of liability in the use of geographic information systems and geographic data sets along with several examples that illustrate theories and limits of liability exposure within geographic information contexts, see Onsrud (1999). The core legal concepts change only slowly over time, so much of the discussion remains relevant. Another primer on liability and geographic information may be found in Cho (2005). While the core legal principles remain relatively stable, practical liability exposure has increased substantially due to the greatly expanded numbers of day-to-day users of digital mapping and guidance systems, the emergence of new use environments for spatial data, and the imposition by legislatures of much larger penalties being applied to cyber violations over comparable physical world violations.

## 17.2  An Illustrative Data Quality Liability Problem

There are obviously many aspects to spatial data quality. The most frequently cited components include accuracy, precision, consistency and completeness, with each of these components typically assessed in terms of space, time and theme (e.g., spatial accuracy, temporal accuracy, and thematic accuracy). One important aspect of data quality relative to completeness is knowing the legal status of a digital dataset that one may desire to copy, or a digital database from which one may wish to extract. That copying or extraction may be by action of a human or software (e.g., automated data mining). If a human or human-initiated software extracts from a geographic dataset or database from which no legal authority exists to extract, substantial liability exposure is incurred.

Just because one finds a music file readily available on the web, does not mean that legal authority has been granted to download it or incorporate all or part of the file content into a derivative work. The same rule applies to geographic data files.

Further, keep in mind that if you were to steal a CD from a music store, the maximum typical fine might be about $1000. However, if you were to download the same ten songs from the Internet, your liability exposure could be as high as $1.5

million[1] (Lessig, 2004: ch. 12). Thus, in a similar manner, if you use a web mapping service or an automated data mining software program to draw together data from ten sources distributed across the web, the same potential $1.5 million liability exposure arises.

What if one only hosts a web mapping service, but it is users that use the technology to draw their own maps extracting data from those other sites? Is the web mapping host site relieved of liability? Among the relevant legal precedents include the Napster (*A&M Records, Inc. v. Napster, Inc.*, 2000) and the recent Pirate Bay cases. In both instances, the operators of the web sites were found guilty even though they only facilitated the sharing of files by others, and even though many instances of sharing were legal. In the Pirate Bay case, decided on 17 April 2009, each of the four defendants was sentenced to a year in jail and fined roughly $900,000 in damages (Carrier, 2009). So in a legal context, in providing a map service, the question arises not whether the operation of the service provides a valuable public benefit or whether legal uses are being made of the site, but whether this is also an operation for facilitating illegal sharing and use of geographic data files without the explicit permission of the owners of those files.

Quality of data requires completeness, and completeness in the global legal environment requires that you know the legal status of the data that you draw from others and have strong confidence that you have legal authority to use the data in your specific context. By far the greatest legal liability exposure for those pursuing exchange and integration of geographic data in a networked world of interoperable web services and data mining is incurred through the violation, whether intended or unintended, of copyright, database legislation, and similar intellectual property protections. Such laws are in place and being continually strengthened by governments across the world.

## 17.3 The Existing Legal Problem with Internet Wide Geographic Data Sharing

The core problem we are confronting is the acquisition by database and dataset developers of automatic copyright upon creation, whether creators want it or not for their data offerings. In most jurisdictions of the world, if one creates an original doodle, copyright occurs instantaneously in that doodle with no need to register the right and regardless of whether most people would find this instantaneous copyright to be reasonable. Therefore, if someone creates a story, song, image or dataset and places it openly on the web, is it free for anyone to copy without permission? Keep in mind that *common practice, not getting caught,* or *small likelihood of being sued* are not equivalent to having a clear legal right to copy. The lawyerly response to the question is that the answer will depend on answers to several additional questions. As a general proposition, however, there will be some minimal creativity in the vast majority of digital works made accessible through the Internet. If minimal creativity exists, the law assumes one must acquire permission from the

---

[1] Under the 1999 Digital Millennium Copyright Act in the United States, statutory damages for behavior unproven as "willful" are set at "not less than $750 or more than $30,000" per infringement while statutory damages for "willful" behavior are not more than $150,000 per infringement. The Recording Industry Association of America (RIAA) has been very effective in using these very high damage limits in extracting large settlements from individual students and others accused of illegal downloading of music files.

copyright holder to copy, distribute, or display the work or generate derivative products from it.

One might argue that 'data' or 'empirical values' drawn from a database are all legally equivalent to 'facts', and therefore are not protected by copyright. Even if true in a specific jurisdiction in a specific instance, the selection, coordination and arrangement of facts may be protected by copyright. Further, the explicit legal tests for qualifying for, or determining what is protected by copyright vary from jurisdiction to jurisdiction (e.g., protection of sweat of the brow, industriousness, etc.) and many jurisdictions supply protections for datasets and databases that extend well beyond those granted by copyright (e.g., database protection legislation, unfair competition regulations, catalog rules, etc.). In truth, one cannot know for certain whether there exists some minimal creativity in a posted geographic data set (and neither can any lawyer) until the gavel falls in a court of law on a case-by-case basis. Thus, most lawyers across the globe, when asked, will advise their clients that they should always assume that a party will emerge to sue unless explicit permission to use is acquired in every instance of drawing from the materials of others.

Assume that you have just extracted data elements from 42 other geographic datasets in an automated data mining or web mapping exercise. Many of us assume that the vast majority of those other sites probably placed their datasets on the web and are adhering to data format and other interoperability standards so that others might freely benefit from their postings. Yet, the law generally holds that we must *not* make this assumption. By example, one or more of those 42 sites has inevitably posted a license provision that your specific use breaches and the posted contract or license provisions of many of the sites are highly likely to be in conflict with each other. Some of the sites will have no license language or use restrictions posted, but in those instances the intellectual property laws of some nation will apply by default. You are required to meet the requirements of all of the involved sites unless you have explicit permission stating otherwise. Shipping a request to the addresses of those 42 sites asking them for explicit permission to use the data extracted from their web sites is burdensome and nonsensical to the typical user. Most of the recipients of your message might also view the request as silly since they would not have posted the data if they did not want it to be used. In many instances you won't receive responses. One should assume however that there are one or more legal claimants lurking out there if you do copy or extract data without explicit permission from each and every source.

Copyright liability is a strict liability concept and thus no intent to break the law or even knowledge of breaking the law is necessary to be found guilty. Even innocent or accidental infringement may produce liability (*DeAcosta v. Brown*, 1944). For each violating extraction or copying, the potential damages are huge and are having a chilling effect on using the geographic data of others, except perhaps for the largest companies with strong legal staffs willing to fight wars of attrition.

Common sense argues that we need strong copyright protections in order to prevent digital thieves from stealing our geographic data offerings against our will. In the world of paper, the law had established a nuanced balance between the rights of creators to benefit from their original works and the rights of new creators to build from the past contributions of those who came before. This balance no longer exists. There are no longer legal exceptions for uses that do not involve a copy such as loaning a book many times or selling a used CD because virtually every use on the Internet involves a new copy. Fair uses and the right to extract such things as facts are readily negated on the Internet by requiring users to click a license before allowing the downloading of data or software. Laws across the globe are skewing even further the legal landscape that has been disturbed by changes in technology rather than restoring the balance. Changed technological conditions and our legislatures in the cause of protecting us from thieves have shifted us to what Lawrence Lessig refers to as a "permissions culture" in the world of the Internet (Lessig, 2004, ch 10). You now almost always need permission to extend from the work of others and this is a world dominated by lawyers.

As a result of the current lack of ability to reliably and efficiently know whether a data set found on the web is available for use without further permission, we have hundreds of millions of orphaned digital works on the web that are not generating any economic income for creators, and in many instances creators placed these data sets on the web for the purpose of being freely used by others. Yet the law states we must *not* assume that we may use them. As a result we do not use the geographic data and products of others in generating our own new products because we are afraid of the legal liability exposure. Either that, or many of us simply use what we find freely available and become law breakers whether we know it or not. This situation is not conducive to supporting democratic societies. When large segments of society become lawbreakers on a daily basis – whether downloading music or geographic data files, breaking the speed limit, or cheating on tax forms – a culture of general disrespect for the law is created. This is bad for society since we generally want our laws respected and enforced. Individuals should also be aware that if evidence exists of commission of a crime such as illegal downloading, certain civil liberties such as the right to privacy, the right to be free of search and seizure of your computer or the right to Internet access are likely to disappear (Lessig, 2004: ch. 12).

To avoid widespread breaking of the law and the creation of a growing culture of disrespect of the law generally, our societal choices are typically to change the law, enforce the law more aggressively, find ways for our community to live within the bounds of the law, or a combination of these approaches.

## 17.4  A Suggested License Solution

Numerous legislative proposals have been submitted in jurisdictions across the globe to restore the balance between the rights of creators and users under copyright law. To date they have consistently failed to gain passage. As mentioned previously, the trend is actually in the opposite direction as legislatures view the need to catch digital thieves as far more pressing than the nebulous need to support freedom to build on the works of others. Rather than fight the law or in addition to advocating changes in the law, another option is to use the law and technology to create an electronic commons in geographic data that all could openly use.

To deal with the data quality issue of completeness, metadata creation for geographic data needs to become very easy, efficient, and fast and become part and parcel of the ordinary course of doing business, science, and government. Those providing data with location components should not even recognize that they are helping to provide standards compliant metadata. Affirmative licensing to allow others to openly access and use geographic data needs to be embedded into the metadata creation process. Further, the ability to determine whether data accessible through the web is available legally for open access use needs to be determinable by browser software. How can these objectives be efficiently accomplished?

A legal commons has already been created for creative works on the web through the use of Creative Commons (CC) licenses (Creative Commons 1, 2009).

With a few clicks, in less than a minute, one may create iron clad licenses for any of your creative work to make it practically and legally accessible to others. Well over 100 million such open access licenses have already been created and the advanced searches of most major web browsers allow one to restrict web searches to return hits to only those sites with the standard CC license you have specified in the search. The browsers automatically pick up embedded html code indicating that the returned sites contain CC licensed material.

There are several versions of CC licenses offered. Every license helps you "retain your copyright (and) announce that other people's fair use, first sale and free expression rights are not affected by the license." …. "Every license allows licensees (i.e., users), provided they live up to (the license provisions), to copy the work, to distribute the work, to display or perform it publicly, to make digital public performances (e.g., web casting), and to shift the work into another (medium). Every license applies worldwide, lasts for the duration of the work's copyright and is not revocable." … "Every license requires licensees to obtain your permission to do any of the things you choose to restrict."(Creative Commons 2, 2009) Among the optional restrictions or conditions you may choose to impose include requiring attribution, restricting uses to noncommercial purposes, not allowing derivative works, and allowing others to distribute derivative works but only under the condition that those works use a license identical to your license (Creative Commons 3, 2009).

The use of such licenses begs the question of what specific CC license would best fit the mores and traditions of science of the past several hundred years, and which would be best for the geographic information science community? For creative works such as research articles and reports I argue that the attribution license with no further restrictions provides the greatest freedom for all of us to extend from the scholarly advancements of each other. Obtaining credit for their contributions is the primary motivating factor and concern of scholars, researchers and students.

Creating licenses for data sets and databases is much more problematic. Creative Commons licenses apply only to creative works and the existence of *creative expression* is very minimal or questionable in many datasets that have been compiled in standard formats to ease machine processing. Thus, as a general proposition, CC licenses should not be applied to data files and databases. Legal scholars have been struggling with what to recommend for several years, and the Science Commons division of CC has recently recommended the use of provisions adhering to the Science Commons open access database protocol. To ensure that geographic data may be used legally across the web for the widest range of purposes, I recommend use of CC0 i.e. Creative Commons Zero or No Rights Reserved (Creative Commons 4, 2009). This language waives all copyright and database rights to the extent that one may have these rights in any jurisdiction and is the best current option to ensure that data can be used legally for general web mapping and feature services, data mining, copying, and extraction. This approach completely avoids the license stacking problem and the need to resolve conflicting restrictions among licensed data sets and those that have been posted without licenses such that default laws of the jurisdiction apply. If contributors affirmatively waive all rights to the greatest extent possible there are no restrictions to conflict with each other. In order to protect attribution for contributors, other means than a copyright derived right may be utilized.

Although minimizing liability for data quality requires completeness of data that incorporates the legal status of the data, merely knowing the legal status of data or knowing the best license for support of web-wide sharing is insufficient. As stated previously, the open access data license generation process needs to be embedded in the metadata generation and recording process in the ordinary day-to-day course of doing business, science and government, or we will merely continue with the status quo of an inefficient system for contributing and finding geographic data and services.

## 17.5 A Suggested Web-Wide Solution

Most people now gathering geographic data that they view as important for some purpose (or they would not have gathered it) are unable to archive these data sets so that the data (a) will be preserved over the long term, (b) can be found again readily, (c) will retain credit for the collector or creator, and (d) can be used legally by others without asking for further permission. Even if accessible through the web, geographic data appropriate for a specific purpose or germane to a specific geographic location is currently very difficult to find, use and share. Producers and users cannot find each other efficiently, nor agree on terms of use efficiently.

Let us consider the example of volunteered geographic information by scientists, students, hobbyists, and many other average citizens. We argue that the typical person engaged in volunteering geographic information about their projects or communities would like to have a simple, comprehensive, practical, and universal solution for providing increased 'findability' for the data and products she produces, wants a solution that is legally, economically and ethically defensible, and does not want details but just a simple solution implemented through the web. We surmise that contributions to the commons would greatly increase and would be far more useful if (1) the ability to contribute geographic data sets was much easier to do, (2) creators could reliably retain credit and recognition for their contributions, and (3) creators would gain substantial benefits by contributing their geographic data to a commons.

In light of the previous discussion about the need for completeness of geographic data to ensure quality and usefulness, the minimum criteria for a widely used and effective web-based contribution facility we hypothesize should include *all* of the following capabilities:

- The ability of contributors to waive their rights to the greatest extent possible in all jurisdictions using iron clad open access legal language (e.g., use the CC0 language) for each contributed data file in less than a minute.
- Accomplish a permissions request to any other potential current or prior rights holder in the data set in less than five minutes with responses automatically fed to the system without human intervention.
- The ability to create standards-based metadata in less than 10 minutes, situating the data in space and time, and tied through an ontology covering all fields of science. We suggest a user interface process with a sophisticated back end by which the system automatically guesses at the field or specialty area of the contribution based on terms supplied by the contributor when describing the file and terms that may be contained in the contributed file. The system would tie the meanings of the terms used to the most appropriate thesauri but as well allow the user to readily correct the guessed at meanings of the words.
- Automatic long-term archiving and backups.

- Automatic conversion of files to interchange formats to enhance survivability of the information over time.
- Attribution acknowledgement. We suggest that the data contributor be credited in the repository for contributing to the shared resource and be granted the right to use the trademark of the facility as acknowledgement of their contributions. This could link to their contributions in the facility (preferred), or attribution could be provided by automatic lineage tracking in all the interchange formats as portions of the contributed data sets are sliced and diced over time. The second option requires more computational and storage overhead, but would allow contributors to search the web to discover how many times their contribution or significant portions of it had been used by others.
- An efficient peer recommendation system.

All of these would be new benefits not typically provided to the typical person volunteering geographic information to web projects today. The contributor would gain easy off-site management and backups of their contributions, enhanced 'findability' of their data sets and data products and that of others, increased recognition and credibility and enhanced peer review of their contributions.

## 17.6  Conclusion

The world is unlikely to see efficient widespread notice of the rights in geographic data sets that would allow legally defensible sharing of geographic data until appropriate rights transfer language is embedded in widely used metadata creation processes. Further, it is doubtful that widespread metadata documentation of geographic data will be universally achieved in doing the day-to-day tasks of business, government and science until all of the above criteria at a minimum are met by operational facilities and distributed across the web.

Similar to the web itself, not all geographic data creators and collectors will contribute their data to a legal commons environment, but many would. Such a commons would constitute a valuable web-wide resource providing assured legal authorization to copy datasets, extract from databases, provide web mapping and web feature services, and engage in data mining. The legal right to carry out such activities under the current status of web and geospatial technology development and in the current global legal environment, however, is very much in question.

## References

Carrier, M.A. (2009), "BitTorrent: Legal Nightmare or Future Business Model?" *OUPBlog* (April 17<sup>th</sup> 2009) Oxford University Press USA. http://blog.oup.com/2009/04/bittorrent/.
Cho, G. (2005), *Geographic Information Science: Mastering the Legal Issues*. John Wiley and Sons, Inc., Chichester, 440p.
Creative Commons 1 (2009) Creative Commons http://creativecommons.org
Creative Commons 2 (2009) Baseline Rights
    http://wiki.creativecommons.org/Baseline_Rights
Creative Commons 3 (2009) Creative Commons Licenses
    http://creativecommons.org/about/licenses/meet-the-licenses
Creative Commons 4 (2009) CC0 http://creativecommons.org/license/zero/
Lessig, L. (2004), *Free Culture: The Nature and Future of Creativity*. Penguin Press, New York, 345p.

Onsrud, H. (1999), "Liability in the Use of GIS and Geographical Datasets". *In:* P. Longley, M. Goodchild, D. Maguire and D. Rhind (eds.), *Geographical Information Systems: Vol.2 Management Issues and Applications*, John Wiley and Sons, Inc., New York, pp. 643-652.

## Legal cases cited

A&M Records, Inc. v. Napster, Inc., 239 F.3d 1004 (9th Cir. 2001), affirming, 114 F.Supp.2d 896 (N.D. Cal. 2000).
DeAcosta v. Brown, 146 F.2d 408 (1944), cert. denied, 325 U.S. 862 (1945).

# 18

## An Example for a Comprehensive Quality Description - The Area in the Austrian Cadastre

*Gerhard Navratil & Rainer Feucht*

### Abstract

*The quality of existing data sets is influenced by technical, legal, and economic parameters. A narrow approach to the description of data quality cannot reflect this complexity. Thus, a more comprehensive approach is necessary. In this paper we use a very small piece of information stored in the Austrian cadastre, the area of a parcel, to show how such a comprehensive discussion could look. We show the technology used to obtain the area, its legal relevance, and the economic impact. We also show with this example the strategies used in Austria to deal with this uncertainty.*

### 18.1 Introduction

The description of the quality of data sets has more than one dimension. From a technical perspective there are lineage, positional accuracy, attribute accuracy, completeness, logical consistency, and temporal information (Guptill and Morrison, 1995). This may be sufficient if the data set has not been interpreted, e.g. a geo-referenced ortho image or a surface model from aerial laser scanning. Semantic accuracy emerges from a linguistic perspective and adds an additional dimension (Salgé, 1995). A legal perspective is necessary if data are influenced by legal re-strictions or instructions (Navratil, 2004). Finally, economic considerations are necessary. Collection of data is costly and thus data are collected only if they pro-vide sufficient economical benefits (Navratil and Frank, 2005). This is already a complex topic if the specifications remain unchanged. However, data describing socially constructed reality (Searle, 1995) must be adapted to reflect changes in the society. Thus, the quality requirements may change. In this paper we show such a description for a simple case: the area of parcels in the Austrian cadastre.

The remainder of the paper is structured as follows: Section 18.2 briefly de-scribes the Austrian cadastre. In Section 18.3 we describe the methods used to determine the area and specify the resulting quality. In Section 18.4 we then show the legal effects of areas and describe what effects may result from a change in the value of the area. In the next section we estimate the economic importance of the area. In Section 18.6 we finally sum up our findings and draw some conclusions.

## 18.2  The Austrian Cadastre

The Austrian cadastral system was established in the 19th century, and consists of cadastral maps and the associated land register (Navratil, 1998). The cadastral maps are a graphical representation of the parcel boundaries and the contemporary land use. The Austrian land register records registration of title (Dale and McLaughlin, 1988: 24-25; Zevenbergen, 2002: 49-55) and thus documents ownership of land. A parcel identification system administered by the cadastral authority connects the two parts.

The Austrian cadastral system originally provided the data for two important tasks: fair land taxation and the protection of private rights on land. Land tax is based on the possible income produced by the land (Fuhrmann, 2007) and is therefore based on the area of the parcel. The area is derived from the boundary, which exists as two different legal qualities:

- as tax cadastre, and
- as coordinate cadastre.

The difference between these two qualities is the legal protection afforded. A boundary in the tax cadastre was agreed upon by the neighbors, and a document was registered containing not only the coordinates of the boundary points but also the signatures of the land owners. This boundary can be restored easily and, since the document is a contract, there can be no more boundary dispute. Although boundaries in the tax cadastre may also be documented by coordinates, they are lacking signatures and thus legal security.

Where do these boundaries come from? Between 1817 and 1861 the whole Austrian-Hungarian Empire was surveyed and all boundaries were drawn on cadastral maps. These surveys were financed by the government and performed by the K.u.K. Ministry of Finance with the help of the Military Geographic Institute. In 1883 a law was passed for updating the data (Franz Josef I., 1883). Since then any change of boundaries must be documented by a licensed surveyor, the resulting documents are then the basis to change the cadastral maps. The process of documentation must be started by the landowner. The landowner commissions a licensed surveyor to document and process the change in the cadastral boundaries. The licensed surveyor is responsible for the documentation and the legal correctness of the resulting parcels (e.g., parcels must have access to public land and must have a minimal size). However, licensed surveyors can only document undisputed boundaries and cannot resolve disputes. This can only be done by courts. As long as the licensed surveyor did not make a gross error in the documentation process, his responsibility ends when the data are recorded in the cadastral map. The licensed surveyor is typically not involved in later processes using the data.

A problem with this approach is that undetected boundary changes are not documented. Land can be acquired by adverse possession. This is the case if a neighbor uses land without knowing he is not the owner. After 30 years he becomes owner but, because he did not know about the adverse possession, this is only detected when comparing the cadastral data to reality. This is usually only the case during a survey, and thus adverse possession is typically detected accidentally. This is one of the reasons why boundaries in the cadastre may be wrong.

The area of a parcel is derived from the boundary and is also part of the cadastre. The precision of the area is square meter. This is fixed in the decree for surveying (VermV, 1994, § 10 Lit. 2). This precision is neither based on technical considerations nor on user requirements. From a data quality perspective the value specified as the area and the value derived from the boundary should always coincide. This was not possible while using analogue maps to display the boundaries. Regular redrawing of the maps was necessary and small deviations between the boundaries of the two maps were inevitable. Thus, a completely new derivation of the areas would have been necessary. This was not done due to personnel restrictions. These differences were easily detectable after creating the digital cadastral map (DKM). Endeavors to investigate the reasons for such differences and match boundaries and areas are performed, but it will take years to eliminate all discrepancies from the cadastral data. Therefore, these areas may still be incorrect because the boundaries themselves may be imprecise or — in cases of adverse possession — wrong. From a data quality perspective the areas are subject to uncertainty.

Seen as an administrative body, the Austrian cadastre is part of the bureaucracy. The main objective of bureaucracy is to produce reliable, predictable, and unambiguous output (Gajduschek, 2003). Uncertainties in the basic data impede the achievement of this goal and must therefore be eliminated or (at least) reduced. Uncertainty reduction originally deals with information uncertainty and its effect on communication. Its advantage compared to other theories is the testable logic and clear formulation (Bradac, 2001). A communication approach for cadastral systems is useful because, as for any geographic information system, communication of information about the real word is a major issue (Bédard, 1987; Frank, 2000). Uncertainty in the original data leads to uncertainty in the decisions made, and this does not correspond with the goals of bureaucracy. A solution to this dilemma is that administrative bodies assume the used data to be correct. In cases of significant errors the administrative body will compensate the affected citizen. This allows a separation between different fields of expertise. In the case of a sale, for example, lawyers or notaries will provide the necessary legal document and use the data from the land register for this. They need to know nothing about the processes used for creating the area values. A licensed surveyor, on the other hand, can assume the correctness of the registered owners when surveying the boundary between two parcels and asking for the owner's approval. This way of dealing with uncertainty is called *institutional uncertainty absorption* (Hunter and Goodchild, 1993).

## 18.3  Methods for Measuring Area

Determining the area within a bounding line is a standard problem. Thus, there are numerous methods to obtain a result. In this section we first show the methods originally used in the 19th century and then move on to the methods currently used. Each of these has an optimum quality, that is, the best possible quality for the method. Additionally there may be relative quality measures. These measures are usually based on the difference between two independent determinations of the area. A comparison with a limiting value (derived from the possible quality) shows if the results are acceptable or not.

### 18.3.1 Methods originally used

In the 19th century obtaining accurate distance measurement was difficult. There-fore, the determination of boundary points was done by angular measurement. The equipment used was the measurement table (Bavir, 1968; Lego, 1968; Zimova *et al.*, 2006), a typical tool at that time (compare, for example, Cousins, 2001). The result is a map showing the boundaries in a specific scale. Typical scales in Austria were 1:720, 1:1440, 1:2880, and 1:5760. 1:2880 was the standard scale and defined as ten times the scale of the topographic mapping (1:28000). Topographic maps were drawn in such a scale that 1 inch on the map represented 1000 paces in reality (12 inches = 1 foot, 6 feet = 1 klafter, 4 klafter = 10 paces). After changing to the international metric system, the maps were rescaled to 1:1,000 and 1:2,000.

The original determination methods for the areas are based on these graphical representations. The following methods were used in Austria as predecessors of the current system:

- Splitting the parcel into shapes with a simple area determination: trian-gles, rectangles, trapezoids, etc. This method is difficult if the bounda-ries are irregular (e.g., along a river or a lake).
- Weighting: The parcel is cut from paper with a specific weight and then compared to other cut-out shapes of which the area is known (e.g., squares).
- Intersect the parcel with a regular grid: The grid cells within the parcel are counted fully and the ones intersected by the boundary line are counted in fractions.

A simplified version of the first method, using triangles only, was used for the current system (Feucht and Navratil, 2004). Necessary measurements were either taken in the field with a tape measure or from the map itself. Checks were per-formed by determining the size of larger areas and comparing with the sum of the corresponding parcels. This check was called group determination. Each parcel was additionally checked by two independent determinations of the area. This check was called single determination because it only affected a single parcel. The maxi-mum allowed differences are listed in Tables 18.1 and 18.2. In both cases the max-imum allowed error is based on the area itself. In case of the single determination, the mapping scale had to be considered (K.u.K. Ministry of Finance, 1904). A detailed discussion of different methods and their accuracy can be found in the master's thesis of Rainer Feucht (2008). An important commonality of all methods is that the quality of the area estimation is worst than a square meter. Thus the li-mited precision does not reduce the quality of the result.

**Table 18.1.** Maximum errors for single determination

| Scale | Max. allowed error |
|-------|--------------------|
| 1:2880 | $dF = 0.001 * F + 0.500 * \sqrt{F}$ |
| 1:1440 | $dF = 0.001 * F + 0.250 * \sqrt{F}$ |
| 1:720 | $dF = 0.001 * F + 0.125 * \sqrt{F}$ |
| 1:2000 | $dF = 0.001 * F + 0.400 * \sqrt{F}$ |
| 1:1000 | $dF = 0.001 * F + 0.200 * \sqrt{F}$ |

**Table 18.2.** Maximum errors for group determination

| Average area of the parcels in the group | Max. allowed error |
|---|---|
| $F \geq 1$ ha | 0.8 * dF |
| 1 ha > $F \geq 0.5$ ha | 0.9 * dF |
| 0.5 ha > F | 1.0 * dF |

In 1969 the decree for surveying (VermV, 1994) specified new error limits for area determination. The new formula was:

$$dF_g = \frac{M}{5000} \sqrt{F}$$

for areas determined from graphical representations. As numerical determination became more important, a different error limit was necessary to compare numerical results with areas derived from the graphical representation. This formula for the maximum difference between graphically and numerically determined areas was:

$$dF_{g-n} = \frac{M}{2500} \sqrt{F}$$

In both cases $F$ is the area of the parcel and $M$ is the scale number (e.g., 1000 for a scale of 1:1000). It can be easily checked whether the allowed error exceeds $1m^2$ for parcels with an area of more than $100m^2$. Almost all parcels in Austria exceed this area and thus the square meter precision is sufficient even today.

### 18.3.2 Methods based on coordinates

Boundaries of parcels in the coordinate cadastre are defined by coordinates and therefore, the area can be computed with standard formulae, e.g., the triangle formula by C.F. Gauß (Gesellschaft der Wissenschaften zu Göttingen, 1929, p. 53; Kahmen, 1993):

$$F = \frac{1}{2} \sum_{i=1}^{n} y_i \left( x_{i-1} - x_{i+1} \right)$$

Since the coordinates have limited accuracy, the determined area will have limited accuracy, too. A standard method to determine the accuracy of a derived value is error propagation with the first order Taylor method (Karssenberg and de Jong, 2005):

$$\sigma_f^2 = \sum_{i=1}^{n} \left( \left( \frac{\partial f}{\partial x_i} \right)^2 \cdot \sigma_{x_i}^2 \right) + 2 \cdot \sum_{i,k=1; i \neq k}^{n} \left( \frac{\partial f}{\partial x_i} \frac{\partial f}{\partial x_k} \cdot \sigma_{ik} \right)$$

The estimation of the area accuracy is thus based on the accuracy of the coordinates. They themselves are based on the accuracy of the control points used for the determination of the boundary points. The precise accuracy of these points is not known. However, there are limits for the accuracy of boundary points and this limit can be used to estimate the accuracy of the area computation.

The parcel shape and correlations between the coordinates have a strong influence on the accuracy of the area. A test with a specific cadastral unit resulted in standard deviation of up to 103m$^2$ based on the standard deviation of the coordinates. The parcels in the test area were rather small (up to 150,000m$^2$) resulting in relative errors of up to 13% (Navratil, 2003). In addition, different correlation models lead to different results (Kraus and Ludwig, 1998).

Discretization is a well-known problem in geometry. Subdivision of a parcel typically requires definition of new points on the boundary lines of the parcel. This is not possible without gaps when working on a regular raster. The solution is to draw the lines to the new points. This keeps a valid topology but slightly changes the shape and thus the area. The Austrian cadastre specifies the coordinates with a centimeter-precision. This leads to two kinds of practical problems:

- Subdivision into two equal area parcels is not always possible.
- The sum of all pieces of a parcel may differ from the area of the parcel itself.

Illustrating the first problem is simple. Let us assume that we have a rectangular parcel of 1000m × 10m and we want to subdivide it into two parcels with a width of 5m each. This task is simple. What happens, if the parcel is not 10m but 10.01m wide? We can only split this into two equal-area rectangles if using millimeter precision and this is prohibited. Thus the subdivision is not possible in the Austrian cadastre.

The same task can be used to illustrate the second problem: this time the length is exactly 1000m. We want to subdivide the parcel into three pieces with a width of 10m each. Each parcel is therefore 1000 ÷ 3 meters long. The first two pieces will have a length of 333.33m because we cannot specify the length with higher precision. The last piece has a length of 333.34m. The areas of all three pieces sum up to 333m$^2$. The rounding therefore produces an error of 1m$^2$.

What are the options for a licensed surveyor in such a situation? There is not much he can do if the original parcel is in the coordinate cadastre. He is not allowed to change the area of the original parcel without changing the boundary. The Austrian Administrative Court ('Verwaltungsgerichtshof', VwGH) pinpointed that the area is a function of the boundary and thus a change of the area is not possible without changing the boundary (Administrative Court, 1995). The licensed surveyor is also not allowed to specify the area with a higher precision. Since the sum of the new areas must match the original area, there is only one solution for the case: he specifies a different area for one of the resulting parcels to compensate for the rounding error. The decision which one this shall be is completely arbitrary.

## 18.4  Legal Importance of Areas

The legal importance of term is defined by the use of these terms. Laws first define terms for specific situations, actions, or a status. Ownership, for example, is defined as *"the competence to rule the substance and the use of a thing to one's arbitrariness and to bar anybody else from substance and use"* (ABGB, 1811, § 354, translated). However, this paragraph does not specify how ownership is acquired or what the effects of ownership are. The acquisition of ownership for typical objects like a book, a car, or a mobile phone requires two steps in the Austrian law: the will of both the current owner and the new owner to transfer the ownership and the handover of the object. The will is usually documented by a contract, which can be an oral one. Handing the object over is simple for small things. In the case of a car it is usually the documents of the car and the car keys that are handed over. In the case of land parcels this is not possible. In Austria the inscription in the land register replaces the handover (§§ 423-431 Austrian Civil Law Code). The acquisition of ownership is thus defined in the same law that defines the term. The effects, however, can be defined in completely different laws. Landowners, for example, have to pay land tax (specified in the Land Tax Law) and if necessary they have to provide land for the construction of railway lines (Law on Compulsory Purchase for Railways).

The same is valid for the area of parcels. The processes for the determination of the area are specified in the law for surveying. The values are then used for other purposes, which are defined in other laws. We discuss the use of the area in these laws briefly. We will start, however, with a discussion of the most obvious use of the area: the sale of a parcel.

### 18.4.1 Area values in the case of a purchase

The land register guarantees for the correctness of the registered owner and the cadastral authority — in case of the coordinate cadastre — for the boundary. There is no guarantee for the correctness of the area in either organization. This may be misleading because purchase contracts for parcels typically specify an area for the parcel. This number is copied from the land register. However, the area in the land register is only an annotation and not a registered (and therefore checked) inscription. The buyer has the obligation to inspect the piece of land in reality and the purchase object is this piece of land and not an object defined by a data set in the land register. There is no guarantee that the area specified in the land register is correct. The inspected piece of land is a real object and is thus subject to uncertainties. The entry in the land register defines a socially constructed object (compare Frank, 2001). The step from the real object to the socially constructed objects absorbs the uncertainty of the values. The purchase object is the real object whereas the sale contract refers to the socially constructed object. Therefore, as decided by the Supreme Court ('Oberster Gerichtshof', OGH), the seller does not have a liability for the area of the parcel (Supreme Court, 1986a) and cannot be sued by the buyer if the parcel has a smaller area than specified in the contract. This is even true if the seller knew about the mismatch. The situation does not even change if there is a registered survey of the parcel and the area value in the land register has not yet been updated (Supreme Court, 1955) or if the value was determined by an expert (Supreme Court, 1986b). Significant discrepancies can only occur if the parcel is in the tax cadastre — the differences in the coordinate cadastre are typically rounding errors (compare Section 18.3.2). A possible way out of this dilemma is thus a survey of the parcel by a licensed surveyor and transfer to the coordinate cadastre. However, there are two points to be considered:

- The costs of the survey have to be paid by either the seller or the buyer. These costs will typically at least €1000 but are insignificant compared to the price of the parcel itself.
- There is a 50% chance that the resulting area is bigger than specified in the land register. In this case the seller will have an argument to raise the price of the parcel.

The risk emerging from the second point is the reason why most of the parcels are not surveyed prior to a sale.

### 18.4.2  Hunting Law

Each land owner has the right to hunt on his land. Austria has small parcels (an average of $8100m^2$) and coordination to preserve natural diversity is thus necessary. Large parcels do not have this problem because they may form a natural habitat of their own. Thus, owners of large parcels can apply for the right of autonomous hunt (*Eigenjagd*) where no such coordination is necessary. The land may be separated into different parcels but they have to be connected and must be in unique owner-ship. There are additional specifications, for example, for minimum width of the area. Finally, the area must have a size of at least $1150000m^2$. The test for this last quality is done by the area specified in the cadastre.

The problem with the specified limit is that the law ignores uncertainty. The test fails if $1m^2$ is missing. The assumption of a relative area uncertainty of 0.1% leads to $1150m^2$. Thus problems in the determination of the area may lead to a wrong result during testing for the right of autonomous hunt. The solution to this problem is the uncertainty absorption in the land register. Since the decision on the right to hunt will not be taken by a surveying expert, considering the uncertainty is almost impossible. Thus uncertainty absorption is a necessary step to simplify bureaucracy.

What happens to a granted right of autonomous hunt if the size falls below the limit after a reevaluation by the cadastral authority? Theoretically, a size reduction should lead to a loss of right. However, it could also be possible that more evidence than just a reevaluation is necessary for such an action. It could require a testimony by an expert since the beneficiary of the right could argue that the correctness of the area is still not guaranteed by the cadastre or the land register. A final answer to this question is not possible because the case has not yet happened. This may be a hint that the Austrian administrative bodies will try to avoid any situation where this problem could occur.

### 18.4.3  Building Law

Austria has nine different building laws, as responsibility lies at the local government level. Thus, there is no unique way that the parcel area is used in the law. However, there are some commonalities:

- Parcels for buildings have to have a minimum size. Typically, for parcels with an area of less than $500m^2$ a building permit will not be issued.
- Spatial planning not only specifies where buildings are permitted, but also defines densities for the buildings. This is either done by specifying the percentage of built areas (the building may cover 80% of the parcel) or by limiting the ratio between the parcel area and the summed up area of all storeys of the building.
- Local taxes must be paid for new parcels for building to compensate for the creation of necessary infrastructure like streets and water supply. These local taxes are typically based on the size of the parcel.

All these limits suffer from the same drawback as the limit in the hunting law. Uncertainties in the estimation of the area are ignored.

### 18.4.4 Forest Law

The definition of forest is based on the use of the area for forestry. This includes areas without trees that are necessary for forestry, like forest roads. Still, the areas must have at least an area of 1000m$^2$ and a width of 10m. This area may be structured in several parcels but for these parcels there is also a minimum size. This size is specified by local laws and is, for example, in Lower Austria 100m$^2$ (specified in the Execution Law for Forestation in Lower Austria).

### 18.4.5 Tax Law

Land tax is based on the potential value produced by the parcel. Factors like soil quality, direction of slope, steepness, etc. influence the productivity of the ground and must therefore be included in the estimation of the value. This is done by categorization. The ministry of finance defines a number of quality classes with reference areas and value numbers ($VN$). During an assessment the parcel is assigned to a specific class. The area of the parcel is only used after this assignment. The productivity ($P$) of a specific parcel is then:

$$P = VN \cdot A$$

$A$ is the area of the parcel. The value for the area is rounded to full 100m$^2$. This is a mismatch to the precision of 1m$^2$ as used in the land register.

It may happen that the cadastral authority performs a correction of the parcel area and this leads to a change of the productivity. Then the land tax will also change. There is no legal problem if the boundary was changed, too. However, if the boundary remained unchanged then the correction eliminated a data error. Unfortunately, the tax prescription is a verdict and the time limit for appeal typically elapsed before the area change is noticed. Thus the overpaid taxes cannot be claimed back.

## 18.5 Economic relevance of areas

It is necessary to know the users when assessing the economic relevance. After this first step we can assess the impact of errors in the parcel size of the outcome of the process (Frank, 2008).

Sometimes the area is used directly. The test for the right of autonomous hunt, for example, directly uses the value. The result is a valuable attribute since such a hunt can be leased. In 2002 the average annual lease was €30 per 10000m$^2$ or at least €3450 per year.

The parcel area is used less directly in several other sections of the administration:

- valuation,
- farming support, and
- assessment of land tax and other taxes.

In Austria three different methods are allowed for valuation: the comparison method uses similar parcels with a known value; the profit method infers the value of the parcel from an assessment of the possible profit; and the object method assesses the value of each part separately (i.e., the parcel itself and all buildings and other objects attached to the parcel). The object method cannot solve the problem of assessing the value of the parcel itself. The other two methods must use the size of the parcel to assess the value. The comparison method, for example, must adapt the values of the compared parcels if the sizes are different.

Farming support could use the areas specified in the cadastre. However, this is a problem if done throughout the European Union because the quality of the area determination may vary. Thus a new determination is necessary. In Austria this is based on ortho images. The farmers mark the extent of their fields and the system can check for overlaps to avoid errors (Sindhuber *et al.*, 2004). Thus, although the area is required, the value from the cadastre is not used.

The assessment of land tax and other taxes is based on the productivity. Land tax ($LT$) is stipulated by the local government:

$$LT = P \cdot TV \cdot RF$$

The tax value ($TV$) may be between 0.05 and 0.2%. Because the productivity excludes improvements like the use of fertilizers, this would lead to a ridiculously low tax. Therefore, a raise factor ($RF$) was introduced, which may be up to 500%. The values for the tax value and the raise factor are defined by the local administration. However, the main point of uncertainty in this process is the assessment of the productivity. The estimation of the value number is rough and has a much higher impact than an error in the area. This is boosted by the fact that the area is rounded to $100m^2$.

The purchase of a parcel may or may not be influenced by the area specified in the cadastre. There is an influence if the purchase price is based on valuation. However, in many cases the price is stipulated by the land owner. A person will buy the parcel if the price seems to be acceptable. A wrong area value cannot have an impact on the price (Supreme Court, 1986a) as discussed in Section 18.4.

## 18.6  Summary and Conclusions

In this paper we discussed the quality of area from different perspectives. The estimation of area from a boundary description is a technical process and has limited accuracy. The use of such a value in a legally defined process ignores this fact and assumes the values to be free of uncertainty. The resulting legal consequences may have economic effects. In some cases the effects of errors in the area may be absorbed by legal rules (e.g., when purchasing a parcel). In most cases, however, the effects are simply ignored. This may lead to interesting legal questions if the area values are corrected:

- Does a decreasing land tax lead to a legally valid claim for repayment of taxes for previous years?
- What happens to a granted right of autonomous hunt if the size falls below the limit?

These cases may cause disputes because they are not unambiguously decided by laws. The Supreme Court decisions and the Administrative Court findings mentioned in the text are all precedents available in Austria. Unfortunately, most similar cases do not make it to the Supreme Court because the decisions of lower courts are accepted or the dispute is solved otherwise. Such solutions do not produce generally valid legal rules because the Austrian law only uses decisions by the Supreme Court and findings by the Administrative and the Constitutional Court as precedents.

It became evident in the discussion that the precision of the area values in the cadastre is problematic in several ways. The accuracy is lower than the precision from a technical perspective. The tools cannot guarantee the correctness of the last digit for large parcels or parcels with complex boundaries. A mismatch can also be found in the tax assessment because the area is rounded to $100m^2$ before it is used. Tax assessment was one of the original purposes of the cadastre. The rounding is therefore a reaction to the uncertainty and shows that the problem of area uncertainty has been detected and resolved.

The simple example of cadastral areas also showed the complexity of such a discussion. The technical processes to determine a value may be simple, the legal consequences, however, may be difficult to determine. One of the reasons for this is that it is often not clear where a specific value might be used. The use of the value may have huge legal and economic effects. Discussing such issues in an international context is difficult because the legal effects will vary between countries. While it might be a logical step to sue a seller if the parcel is smaller than specified in the sale contract, this is not possible in the Austrian jurisdiction. Similar examples can be found in other areas. All experts are trained in a specific jurisdiction and this may influence the treatment of data quality questions. Discussion within the limited scope of a single jurisdiction may lead to difficulties when trying to assess general effects of data quality.

## Acknowledgements

We thank Julius Ernst and Christoph Twaroch for their contributions during discussions of the topic. The comments of three anonymous reviewers were helpful for significant improvement of the paper.

## References

Administrative Court (1995), Finding. *VwGH 94/06/0026*.
ABGB (1811) Austrian Civil Law Code (Allgemeines bürgerliches Gesetzbuch), *JGS Nr. 946/1811*.
Bavir, A. (1968), *Skriptum zur Vorlesung Vermessungskunde*. Vienna University of Technology, Austria.
Bédard, Y. (1987), "Uncertainties in Land Information Systems Databases". *In:* N.R. Chrisman (ed.) *Proceedings of the Autocarto 8*, Baltimore, Maryland, pp. 175-184.
Bradac, J.J. (2001), "Theory Comparison: Uncertainty Reduction, Problematic Integration, Uncertainty Management, and Other Curious Constructs". *Journal of Communication*, Vol. 51(3): 456-476.
Cousins, S.A.O. (2001), "Analysis of land-cover transitions based on 17th and 18th century cadastral maps and aerial photographs". *Landscape Ecology*, Vol. 16: 41-54.

Dale, P.F. and J.D. McLaughlin (1988), *Land Information Management. An introduction with special reference to cadastral problems in Third World countries*. Oxford, Oxford University Press, U.K., 266p.

Feucht, R. (2008), *Flächenangaben im österreichischen Kataster*. Masters thesis, Vienna University of Technology. Austria, 93p.

Feucht, R. and G. Navratil (2004), "Flächenangaben im Kataster aus historischer Sicht". *In:* J. Strobl, T. Blaschke and G. Griesebner (eds.). *Angewandte Geoinformatik 2004, Proceedings of the AGIT*, Salzburg, Austria, pp. 123-128.

Frank, A.U. (2000), "Communication with maps: A formalized model". *In:* C. Freksa, W. Brauer, C. Habel and K.F. Wender (eds.). *Spatial Cognition II (Int. Workshop on Maps and Diagrammatical Representations of the Environment, Hamburg, August 1999)*. Berlin Heidelberg, Springer-Verlag. Vol. 1849: 80-99.

Frank, A.U. (2001), "Tiers of ontology and consistency constraints in geographic information systems". *International Journal of Geographical Information Science*, Vol. 75(5): 667-678.

Frank, A.U. (2008), "Analysis of Dependence of Decision Quality on Data Quality". *Journal of Geographical Systems*, Vol. 10(1): 71-88.

Franz Josef I. (1883), *Gesetz über die Evidenthaltung des Grundsteuerkatasters* (Law on updating the cadastre). *RGBl.Nr 1883/83*: 249-268.

Fuhrmann, S. (2007), "Digitale Historische Geobasisdaten im Bundesamt für Eich- und Vermessungswesen (BEV) - Die Urmappe des Franziszeischen Kataster". *Österreichische Zeitschrift für Vermessung & Geoinformation*, Vol. 95(1): 24-35.

Gajduschek, G. (2003), "Bureaucracy: Is it Efficient? Is it not? Is that the Question?". *Administration & Society*, Vol. 34(6): 700-723.

Gesellschaft der Wissenschaften zu Göttingen (1929), *Carl Friedrich Gauß Werke*, Vol.12, Berlin, Julius Springer, Germany, 415p.

Guptill, S.C. and J.L. Morrison, (eds.) (1995), *Elements of Spatial Data Quality*, Elsevier Science, on behalf of the International Cartographic Association, U.K., 202p.

Hunter, G.J. and M.F. Goodchild (1993), "Managing Uncertainty in Spatial Databases: Putting Theory into Practice". *URISA Journal*, Vol. 5(2): 55-62.

K.u.K. Ministry of Finance (1904), *Instruction zur Ausführung der trigonometrischen und polygonometrischen Vermessungen behufs neuer Pläne für die Zwecke des Grundsteuerkatasters*. Wien, K.u.K. Hof- und Staatsdruckerei, Austria, 233p.

Kahmen, H. (1993), *Vermessungskunde*. Berlin, de Gruyter, Germany.

Karssenberg, D. and K. de Jong (2005), "Dynamic Environmental Modelling in GIS: 2. Modelling Error Propagation". *International Journal of Geographic Information Science*, Vol. 19(6): 623-637.

Kraus, K. and M. Ludwig (1998), "Genauigkeit der Verschneidung von geometrischen Geodaten". *Zeitschrift für Vermessungswesen*, Vol. 123(3): 81-87.

Lego, K. (1968), *Geschichte des österreichischen Grundkatasters*. Vienna, Bundesamt f. Eich- und Vermessungswesen, Austria, 76p.

Navratil, G. (1998), *An object-oriented model of a cadaster*. Master thesis, Vienna University of Technology, Austria, 91p.

Navratil, G. (2003), "Precision of Area Computation". *In:* ESRI Geoinformatik GmbH (ed.) *Proceedings of the ESRI 2003 - 18. European User Conference*, Innsbruck, Austria.

Navratil, G. (2004), "How Laws affect Data Quality". *In:* A.U. Frank and E. Grum (eds.) *Proceedings of the International Symposium on Spatial Data Quality (ISSDQ)*, Bruck a.d. Leitha, Austria, Department of Geoinformation and Cartography, pp. 37-47.

Navratil, G. and A.U. Frank (2005), "Influences Affecting Data Quality". *In:* L. Wu, W. Shi, Y. Fang and Q. Tong (eds.) *Proceedings of the International Symposium on Spatial Data Quality (ISSDQ)*, Beijing, China, pp. 234-242.

VermV (1994), Austrian Decree for Surveying (Vermessungsverordnung), *BGBl. Nr. 562/1994*.

Salgé, F. (1995), "Semantic Accuracy". *In:* S.C. Guptill and J.L. Morrison (eds.) *Elements of Spatial Data Quality*. Oxford, Elsevier: pp. 139-151.

Searle, J.R. (1995), *The Construction of Social Reality*. New York, The Free Press, U.S.A., 240p.

Sindhuber, A., M. Bauer, C. Golias, T. Nemec, M. Ratzinger, G. Rauscher and T. Weihs (2004), *"INVEKOS-GIS - Ein Internet-GIS für Landwirte"*. *In:* J. Strobl, T. Blaschke and G. Griesebner (eds.). *Angewandte Geoinformatik 2004, Proceedings of the AGIT*, Salzburg, Austria, Wichmann, pp. 632-637.

Supreme Court (1955), Decision. *OGH 1Ob272/55*.

Supreme Court (1986a), Decision. *OGH 8Ob654/86*.

Supreme Court (1986b), Decision. *OGH 1Ob679/86*.

Zevenbergen, J. (2002), *Systems of Land Registration Aspects and Effects*. Delft, Netherlands Geodetic Commission, The Netherlands, 210p.

Zimova, R., J. Pestak and B. Veverka (2006), "Historical Military Mapping of the Czech Lands – Cartographic Analysis". *In:* T. Bandrova (ed.) *Proceedings of the 1st International Conference on Cartography and GIS*, Borovets, Bulgaria, 7 p.

# 19

## Data Protection, Privacy and Spatial Data

*Teresa Scassa & Lisa M. Campbell*

### Abstract

*In this paper, the authors explore the extent to which spatial data may be considered personal information for the purposes of data protection and privacy law. While data quality is an important objective in the creation of spatial data applications, the authors demonstrate that even relatively low quality spatial data may attract the application of data protection or privacy law, particularly when it is matched or combined with other data sets. The rapid development of a variety of applications and tools that incorporate spatial data pose significant privacy law challenges both for individuals and for the developers and users of these tools.*

### 19.1  Introduction

Enhancing spatial data quality is typically seen as a virtue; something that will improve the overall accuracy and reliability of the data for a variety of purposes. However, accurate spatial data can raise serious privacy concerns. Accurate spatial data, particularly when matched with other data sets, can often lead to the identification of individuals, and to the association of a range of data with those individuals. The widespread and increasing public availability of an almost infinite variety of data sets means that data can be matched and analyzed with other data sources in order to identify individuals using highly significant geographic markers. In this context, even vague or imprecise geographical information, when matched with other data, may lead to the identification of specific individuals. The potential for identification of individuals through the linking of spatial data with other data poses challenges for access to information regimes, public health and medical researchers, and for all those who must comply with data protection legislation.

In this paper we consider several situations where spatial data can give rise to privacy problems when it is combined with other data sets. These include online maps with three-dimensional imagery such as Google Street View and Microsoft Virtual Earth, neighbourhood crime maps furnished by police services, and the matching of census data or other data to postal code information. The examples serve only as illustrations of a complex and wide reaching problem, and through them we seek to identify some of the normative challenges that arise. We also consider the impact of technologies such as mobile devices and molecular computing.

## 19.2   Personal Information

Spatial data on its own is typically not personal information. However, spatial data when matched with other data about individuals may become personal information. In some cases, the spatial data may be the key to identifying the individual referenced in other sets of data. Data protection regimes typically protect 'personal information'. Thus it is important to understand the point at which spatial data may become personal information. In Canada most data protection and access to information legislation defines 'personal information' as being, at its essence, information *"about an identifiable individual"* (*PIPEDA*, § 2; *Privacy Act*, § 3). Courts have taken a fairly broad approach to interpreting this concept. In the access to information context, they have stated that information is personal information if it is 'about' an individual and if it permits or leads to the possible identification of the individual" (*Canada (Inf. Comm'r) v. Canada (Transport Accident)*, 2007). This possible identification might occur through linking the data with *"information from sources "otherwise available" including sources publicly available"* (*Gordon v. Canada*, 2008: ¶ 33).

The standard for 'identifiability' set by the Federal Court of Canada in *Gordon v. Canada (Minister of Health)* was that of a *"serious possibility"* (*Gordon v. Canada*, 2008: ¶ 34). Using a standard such as serious possibility offers a very broad scope to the term 'personal information' and thus to the legislation. A serious possibility of identification does not necessarily mean actual identification; nor does it require that all the relevant data be available in one place or from one source. However, there is no broad national consensus as to the standard to apply, and *Gordon* dealt with the concept of privacy in access to information legislation. Under Ontario's *Freedom of Information and Protection of Privacy Act* (1990), identifiability has been assessed on the basis of whether there is *"a reasonable expectation that the individual can be identified from the information"* (*Ontario v. Pascoe*, 2002: ¶ 1). Under such a standard, a decision-maker's view of the reasonableness of any expectation may be influenced by the sensitivity of the information, or by the cost or labour involved in the matching process (*Minister of Community Safety*, 2008). Further, under the Ontario statute, it may be that some evidence demonstrating identifiability may be required. In other words, it may be necessary to show how taking certain steps and matching certain data will result in the identification of at least one individual (*Minister of Community Safety*, 2008). These divergent standards (serious possibility v. reasonable expectation) increase the difficulty in determining when information is likely to be considered 'personal information'. The answer will depend in part upon the particular legislative regime at issue in any given case, and in part on highly context-specific facts. Nevertheless, under both the *Gordon* and *Pascoe* tests, the question of whether certain data is 'personal information' is not answered by considering that data in a vacuum. Data held by other parties, and even data sets generated after the collection of the data at issue, may render that information 'personal information' if by matching those data sets with the data in question, an individual may be identifiable.

It is important to note that the assessment is always contextual. Whether spatial data is personal information will depend upon the amount, nature and quality of any other data with which it might be matched. Other facts may also be relevant including the ease with which matching can take place, the accessibility of other data sets, and the degree of harm that might flow from any identification.

### 19.2.1 Postal Code Data

Postal code data is a form of geospatial data that is less precise than a specific address or set of geographic co-ordinates, but that nevertheless raises significant privacy concerns in a variety of contexts. These contexts include the use of data in medical and public health research, direct marketing, academic study, and many other forms of research that rely upon census-based demographic data sorted according to postal code and matched with other data sets. Because a postal code may be assigned to a single apartment building or to five single homes in an urban environment (*Minister of Community Safety*, 2008), a postal code combined with other data may effectively identify specific individuals. In Canada, Statistics Canada sells a Postal Code Conversion File that permits the linking of six digit postal codes with census demographic data. The linking of postal code data with census data provides an already data-rich set of linkages that may, independently or when matched with other data, permit the identification of individuals, even though Statistics Canada does take steps to protect confidentiality through random rounding and other practices. In the US, Sweeney has established that a five digit zip code combined with data about gender and date of birth will uniquely identify 87% of the US population (Sweeney, 2002). In the UK marketers have found that *"a link via the postcode system with the electoral roll means that it is possible to identify individual households and their characteristics"* (Evans, 2005: 106).

Postal code data is not the most precise form of spatial data, yet even less precise geographic indicators may lead to information being characterized as 'personal information'. In *Gordon v. Canada (Minister of Health)* (2008), a journalist was denied full access to data sought from the Canadian Adverse Drug Reactions Information System (CADRIS). In particular, the Federal Court ruled that data about the province in which an adverse drug reaction had been reported must not be disclosed because to add that data to the other data being disclosed:

> ". . . would substantially increase the possibility that information about an identifiable individual that is recorded in any form would fall into the hands of persons seeking to use the totality of information disclosed from the CADRIS database, in conjunction with other publicly available information, to identify "particular" individuals" (*Gordon v. Canada*, 2008: ¶ 43).

It is important to note that the court placed no limits on the other information with which the province of residence data might be matched. In an age where more and more data is being made publicly available, in free or in commercial contexts, the potential for identification or re-identification must be constantly increasing. It is also worth noting that in *Gordon* the court seems to accept that the matching information might be that held by a co-worker or a neighbour (*Gordon v. Canada*, 2008: ¶.43). Thus it may be that identifiability must be considered in terms of both formal and informal sources of information.

Postal code data, when matched with geodemographic data, such as that made available by Statistics Canada (Statistics Canada, 2006), is of keen interest to marketers who seek increasingly accurate ways to target their product marketing to specific demographic groups. Although the mapping of postal codes to geographic areas, and thence to geodemographic information does not necessarily reveal personal information about specific individuals, interesting questions are raised when this data is further matched to specific individuals. For example, if geodemographic assumptions about individuals living at a particular postal code are then matched to the individuals known to use that specific postal code, can it be said that the information is about an identifiable individual? There is case law to support the view that information does not need to be true or accurate to be personal information, so long as it is associated with an identifiable individual (*Lawson v. Accusearch Inc.*, 2007). If this is the case, generalized geodemographic information that is linked to a specific individual through the vehicle of a postal code is personal information, even if that individual does not fit the geodemographic profile that has been assigned to them. The implications of this could be significant, as practices of this kind underpin many direct marketing initiatives that would, if this were the case, be non-compliant with data protection laws. The example illustrates another way in which relatively imprecise spatial data, combined with generalized geodemographic information, can become personal information about identifiable individuals, with the attendant legal privacy consequences.

## 19.2.2  Three-dimensional maps online

Satellite and aerial images of urban centres have been available on the World Wide Web for some time. Generally, these are overhead images of large tracts of land, and of moderate resolution such that it is impossible to discern images of individuals in the area. That dynamic has changed with recent developments in three-dimensional mapping technology.

Using a variety of techniques, several companies have created online maps that display street-level images at high resolution. Photosynth[1] gathers regular two-dimensional photographs of a given area and combines them into a 3D image. Everyscape[2] also turns 2D images into 3D, and invites surfers to submit images to add to the database. Microsoft's Virtual Earth[3] integrates aerial and 3D imagery with mapping and search functions, and Google's Street View[4] allows viewers to navigate within street-level imagery that was captured previously. Earthmine, another 3-D mapping company, has cars equipped with cameras that employ a stereoscopic system with wide angle lenses that capture spherical images of the surrounding environment.[5]

These images show buildings, cars and people in the area at the time that the image was captured, and they are enriched with mapping and search functions. The creators of online maps seek to give users an experience akin to a souped-up version of walking along a street. The user will see images of the street, its build-

---

[1] http://livelabs.com/photosynth

[2] http://www.everyscape.com

[3] http://www.microsoft.com/virtualearth

[4] http://maps.google.com/help/maps/streetview

[5] http://www.earthmine.com

ings, attractions and people present there at the time the images were captured. Users may also gather and share information about businesses, restaurants and other attractions.

Hundreds of cities around the world now appear in 3D online maps, and contained in them are images of countless individuals. To the extent that individuals may be identified from the images, the images will be 'personal information', thus bringing them within the scope of data protection legislation. Generally speaking, these images appear to be captured without the knowledge and consent of the individuals featured in them. Data protection legislation typically requires consent for the collection, use or disclosure of any personal information. Partly in response to public pressure, some companies producing these maps have taken steps to anonymize personal information, either by blurring faces and other identifying features, or by using lower resolution technology so that personal information cannot be detected. However, the reality is that it is cheaper and faster to capture high resolution data with rich personal information than it is to take steps to protect privacy, such as blurring and anonymization.

As noted earlier, under Canadian data protection legislation, any information may be personal information if it can be linked to an identifiable individual either on its own or using other data sources. This raises questions about the initial collection of the information by street level imaging services, as well as use and distribution. For example, it remains to be seen whether the blurring of faces and other identifying details such as license plates is adequate in all cases to eliminate the identifiability of individuals. It is also not clear that faces and license plate numbers constitute the only information likely to be considered 'personal information' in these applications. If other information in the images can be linked to identifiable individuals using other available sources, then it will likely be considered personal information. According to a recent news story related to the impending launch of Google's Street View in Canada, Google will automatically blur faces and licence plates on images that are publicly released. However, they plan to retain versions of the original unblurred images (CBC, 2009). Not only will these original images in some cases constitute personal information, the original images may themselves be the 'other available information' that will make the blurred images identifiable.

### 19.2.3  City crime statistics

The City of Ottawa, Ontario, recently launched a Google-based crime-mapping tool that allows the public to view police call data (Ottawa Crime[6]). Although Ottawa is the first major Canadian city to implement such a program, similar tools are available in many other cities worldwide. The data included in the maps is based on location of calls for police assistance, and the calls are grouped by categories such as theft, assault and homicide. A number of crimes are not featured on the map due to privacy concerns, including spousal and sexual assaults.

The service is free of charge, and users may select specific time frames as well as precise districts within the city, including different geographic boundaries, city wards, police districts and patrol zones. Data is refreshed daily, and by clicking on icons, users may see the time and date of a call along with its reference number.

---

[6] http://www.ottawapolice.ca/en/resources/crime_analysis_statistics/crimereports.cfm

Tools such as Ottawa Crime are meant to fill an important role in providing information of interest to the public in choosing neighbourhoods, or in being aware of risks and dangers near their homes or workplaces. Nevertheless, as US-based researchers have noted, such tools must balance the broader public interest in access to this form of information with the public interest in the protection of privacy (Wartell and McEwen, 2001). Privacy can be compromised when crime mapping tools provide overly detailed information – including precise location data. As a result, privacy protective features are required. In the case of Ottawa Crime, for example, the names of individuals associated with the events are not included. There is no indication of whether a call led to criminal charges being laid, or whether there was a conviction. Significantly, street names are available, although not actual house numbers. However, data is mapped at 100 block increments. It is not clear yet whether enough has been done to render this tool compliant with data protection norms. As a municipal government initiative, it would fall under the *Municipal Freedom of Information and Protection of Privacy Act* of Ontario, and thus under the standard used in Ontario data protection law – that of the 'reasonable expectation' of identifiability. If it can be demonstrated that a particular individual – for example, a victim of a crime – can be identified using the data from this tool in conjunction with data from other sources, this standard would likely be met, and the tool would be non-compliant.

These issues are not new. In a 2001 discussion paper, the United States Department of Justice Crime Mapping Research Center studied the problems associated with posting crime data (Wartell and McEwen, 2001). These include protecting victims' privacy interests, ensuring accuracy and controlled access to crime data, and potential misuse of the data. While the authors offered some guidance to law enforcement agencies interested in publishing and sharing crime data, they concluded that there were many unresolved issues related to crime mapping and a need for further exploration and analysis. Among other privacy issues, the report identified concerns that secondary uses of the data might be made by private sector companies. For example, insurance companies might make decisions about rates or insurability based on the neighbourhood crime data, and companies selling alarms or other security features might target specific areas for marketing campaigns. While one could argue that this targeting makes use of information about a neighbourhood and not specific individuals, the association of specific individuals with particular neighbourhoods and with assumptions about those neighbourhoods may, in fact, raise privacy and other concerns (Gandy, 2006: 378). In the Canadian data protection context it should be noted that legislation typically requires that information collected for specific purposes should not be used for other purposes without consent.

It is interesting to note that the modification to the data required to protect individual privacy may also present data quality issues. There is a risk that a distorted public perception may flow from partial or inconsistent reporting of crime statistics, or from the flawed manner in which they are represented (Wartell and McEwen, 2001: 18).

### 19.2.4  Mobile devices and molecular computing

Three-dimensional online maps are migrating towards mobile devices, including Google's Android and Apple's iPhone. Other companies are also developing technologies for this emerging market. For example, using global positioning systems, Enkin's handheld navigation determines a user's position while motion sensors calculate the device's orientation in space, and a combination of 3D graphics and live video are displayed to the user.

Google introduced Google Latitude[7] early in 2009. Latitude is a location-sharing service that allows users to share location data with others, either using a mobile phone or a Google Gears-equipped computer. The service uses a digital map to automatically show where an individual is at a given time, often pinpointing location to within a few metres. Location is established using GPS, and is shared with those the user selects via the Latitude interface.

A month later, Yahoo released a Facebook application called 'Friends on Fire'[8] that allows users to share their location with one another. The application builds on Yahoo's pre-existing Fire Eagle service[9], which stores location information fed to it from mobile devices or from online user input, and distributes that information to third-party services. Installing the Friends on Fire application would allow Facebook users to share location information with other users. Friends on Fire is but one of numerous applications that have been developed using the Fire Eagle service. These applications are part of a growing trend. More companies are creating software to collect and share location data, as this information is widely seen as a way to make local and mobile advertising more effective. Privacy advocates have warned of the dangers of these applications, including a lack of adequate safeguards that lead to a risk of covert surveillance (Privacy International, 2009).

Coupled with these developments in mobile computing is a revolution in computing technology through the development of molecular computing. Over the last 60 years, ever-smaller generations of transistors have driven exponential growth in computing power. Atomic-scale computing, in which computer processes are carried out in a single molecule or using a surface atomic-scale circuit, holds vast promise for the microelectronics industry. It allows computers to continue to increase in processing power through the development of components in the nano- and pico scale. Theoretically, atomic-scale computing could put computers more powerful than today's supercomputers in everyone's pocket.

These devices will ultimately enable their developers to pinpoint with great accuracy the location of individuals using or possessing these devices. This will raise data protection issues in terms of the need for customer consent to the collection and use of this data by the company. It will also likely raise significant privacy issues in terms of the conditions under which this data may be accessed, used by or disclosed in the context of civil litigation, or for the purposes of law enforcement or national security (Nouwt, 2008). Such issues have already arisen in the criminal context with respect to the current ability to locate individuals through cell phone usage (*R. v. Mahmood*, 2008).

Significant privacy concerns are raised by the increasing collection of location data, and by the growing accuracy of such data. Location information about customers that is stored by service providers may be shared in unanticipated ways with other private sector companies under vague or open-ended terms in privacy policies. Location data can be an extremely important component of a data profile, as it has

---

[7] http://www.google.com/latitude/intro.html

[8] http://apps.facebook.com/on-fire

[9] http://fireeagle.yahoo.net

the potential to significantly enhance the value of other information available about the data subject.

Privacy concerns are not just with respect to the use of personal information by private sector actors. Data in the hands of private sector companies may be sought by police or other state authorities for a range of purposes including investigations. Even basic cell phone technology will permit the matching of an individual's general geographic location at the time a call was made with their name and address, and with information about the person they called. The expanding collection of increasingly precise location data about individuals by private sector companies provides state authorities with unprecedented opportunities to retroactively track individuals or to generate lists of suspects based on location at specific places at certain points in time (*R. v. Mahmood*, 2008). Spatial data may thus contribute to data pictures that raise constitutional privacy concerns.

## 19.3  Conclusion

There is a growing recognition in data protection and privacy law that location information may be personal information – either because it is about an identifiable individual or because it provides the key, when linked with other data, to identifying specific individuals. It may also situate an individual at a particular point in space and time, and even provide information about whether the individual is engaged in a particular activity. Thus while high quality spatial data is an objective in many products and applications, some applications may require a degrading of the quality of the spatial data in order to avoid raising privacy issues. Nevertheless, it is important to note that spatial data need not be of particularly high quality for these same concerns to arise; even postal code data may be sufficient, when linked with other data, to raise privacy concerns. Further, the imprecise geographic data matched with other generalized data may result in 'personal information' for legal purposes, even if the resultant information is imprecise or inaccurate.

The ability to match spatial data with other data sets, the growing volume of data that is available for matching, and dramatically increasing processing capabilities create powerful means by which previously 'anonymized' data may become personal information, with attendant legal consequences. In many cases, spatial data may be a key factor in linking other data sets to identifiable individuals. Data protection law suggests that information will be personal information where there is a 'serious possibility' or a 'reasonable expectation' that an individual may be identified using that data and other available data. The uncertainty of the test to be applied and the unpredictable nature of the other data with which spatial data might be matched pose challenges for developers of spatial data applications who must be attentive to privacy and data protection issues.

## Acknowledgements

The authors acknowledge the support of the GEOIDE Network and the Canada Research Chairs Program. The views expressed in this paper are those of the authors alone, and do not necessarily reflect those of the Office of the Privacy Commissioner of Canada.

# References

Canadian Broadcasting Corporation (CBC). (2009), "Google alerts Canadians about Street View filming", March 26<sup>th</sup> 2009, *CBC News: Technology and Science.* http://www.cbc.ca/technology/story/2009/03/26/tech-090326-google-street-view.html.

Evans, M. (2005), "The data-informed marketing model and its social responsibility". *In:* S. Lace (ed.), *The Glass Consumer: Life in a Surveillance Society,* Bristol, The Policy Press, pp. 99-132.

Gandy, O.J. (2006), "Data Mining, Surveillance, and Discrimination". *In:* K.D. Haggerty and R.V. Ericson (eds.), *The New Politics of Surveillance and Visibility,* Toronto, University of Toronto Press, pp. 363-384.

Nouwt, S. (2008), "Reasonable Expectation of Geo-Privacy?". *SCRIPTed,* Vol. 5(2): 375-403. http://www.law.ed.ac.uk/ahrc/script-ed/vol5-2/nouwt.asp

Privacy International. (2009), "Privacy International Identifies major security flaw in Google's global phone tracking system", *Privacy International News Bulletin,* February 5<sup>th</sup> 2009, http://www.privacyinternational.org/article.shtml?cmd[347]=x-347-563567.

Statistics Canada. (2006), *Census Data Products,* http://www12.statcan.ca/census-recensement/2006/dp-pd/index-eng.cfm.

Wartell, J. and T.J. McEwen. (2001), *Privacy in the Information Age: A Guide for Sharing Crime Maps and Spatial Data.* U.S. Department of Justice. http://www.ncjrs.gov/pdffiles1/nij/188739.pdf

# Legal cases cited

*Canada (Information Commissioner) v. Canada (Transport Accident Investigation & Safety Board)* [2007] 1 F.C.R. 203 (F.C.A.).

*Freedom of Information and Protection of Privacy Act (Ontario), R.S.O. 1990, c. F. 31.*

*Gordon v. Canada (Minister of Health),* 2008 FC 258 (CanLII).

*Lawson v. Accusearch Inc.* 2007 F.C. 125.

*Minister of Community Safety and Correctional Services,* Order PO2726, Ontario OIPC, October 22, 2008.

*Municipal Freedom of Information and Protection of Privacy Act (Ontario),* R.S.O. 1990, c. M. 56.

*Ontario (Attorney General) v. Pascoe,* (2002), 22 C.P.R. (4th) 447.

*Personal Information Protection and Electronic Documents Act,* S.C. 2000, c. 5 (PIPEDA).

*Privacy Act,* R.S.C. 1985, c. P-21.

*R. v. Mahmood,* [2008] O.J. No. 3922 (Ont. S.C.).

# 20
## Short Notes

## Harmful Information: Negligence Liability for Incorrect Information

*Jennifer Chandler*

One of the issues facing designers of geographical information systems (GIS) and distributors of spatial data is the possibility of legal responsibility for defects in their systems and data.

There is a relatively small group of legal decisions that pertain directly to liability for defects in spatial data and GIS. These cases address errors in maps and charts, navigational information (e.g., visibility, local weather conditions), defects in positioning instruments and systems (e.g., altimeters), and surveys (see, for example, Phillips, 1998-1999). However, this group of cases, which deals specifically with one form of geographical information or another, is a limited subset of a much broader pool of cases that deal with liability in the context of various types of information.

It is useful for understanding liability in the context of GIS and spatial data to consider the issue of liability for incorrect information in general. It provides a richer and broader group of cases within which to identify the key considerations in understanding liability for defects in GIS and spatial data. At the same time, it will be important to note the ways in which spatial data may differ from other forms of information. These differences will affect the extent to which one can, on the basis of this broader set of information-related court cases, predict how courts will handle disputes involving incorrect spatial data. These differences may be inherent to the type of information, or may stem from the social structures within which spatial data is produced and used (Gahegan and Pike, 2006).

Among the questions that designers and distributors of GIS and spatial data should consider are: (a) what is the legally-required level of care to ensure the safety of spatial data or a GIS (at the design, manufacturing and marketing stages); (b) what forms of caveats should be included with information that may pose dangers; (c) what procedures should be followed when errors or dangers are discovered (e.g., warnings, recalls, corrections); (d) what are the principles by which we assign responsibility among the initial disseminator of information, re-publishers who may pass the information along, value-added users of original information who modify or add to the information, and the end-users who decide to rely upon the information; and (e) what is the extent to which providers of information have been able to limit liability through various means such as contracts and licenses?

However, the utility of the broader category of cases dealing with various kinds of information liability is not limited to its larger number of cases available for consideration. A focus on information liability brings to the fore a variety of other important policy considerations related to information that may be interesting and relevant in the context of geographical information.

Laws regulating information and liability for incorrect information will raise issues that do not come up at all or in the same way with laws governing other aspects of human behaviour. One of these additional issues is the matter of freedom of speech. Freedom of speech considerations have led some courts to resist finding liability for inaccurate information in the context of books (Noah, 1998). It is possible that freedom of speech values may be relevant in some cases with GIS and spatial data.

Information also raises other issues that cause courts to treat it differently in the liability context. For example, certain properties of information, such as the ease of propagation or the economics of production, are different for information compared to other forms of products. These factors tend to affect the judicial reasoning on the appropriate level of responsibility to impose upon those who disseminate information products. Ownership of information is also subject to the specific rules of intellectual property law. To the extent that legal responsibility for inaccurate information rests with those who own and/or commercially exploit information, uncertainty or vagueness in those rules will complicate the identification of responsible parties. In the context of GIS and spatial data, new models of production including open source software development and data mash-ups represent two questions to address.

Among the broad range of information-related cases that we can consider to identify how courts address information liability are cases dealing with numerous types of professional advice or instructions (e.g., medical advice, auditors' reports, engineers' drawings), the publication of facts and instructions in printed products (e.g., books and newspapers), defects in software and computerized systems, etc. In addition to instances where the provision of erroneous information has been harmful, the law recognizes as harmful the failure to provide information in a range of contexts (e.g., disclosure of risks in the medical context, warnings in relation to product hazards). All of these contexts, while not specifically focused on GIS and spatial data, are clearly relevant to practitioners in the field of GIS and spatial data.

# References

Gahegan M. and W. Pike. (2006), "A Situated Knowledge Representation of Geographical Information". *Transactions in GIS*, Vol. 10(5):727-749.

Noah, L. (1998), "Authors, Publishers and Products Liability: Remedies for Defective Information in Books" *Oregon Law Review*, Vol. 77:1203-1228.

Phillips, J.L. (1998-1999), "Information Liability: The Possible Chilling Effect of Tort Claims against Producers of Geographic Information Systems Data" *Florida State University Law Review,* Vol. 26: 756-781.